A DARWINIAN WORLDVIEW

To Jo Elvidge and to the memory of Arnold Elvidge

A Darwinian Worldview
Sociobiology, Environmental Ethics and the Work of Edward O. Wilson

BRIAN BAXTER
The University of Dundee, Scotland, UK

Routledge
Taylor & Francis Group
LONDON AND NEW YORK

First published 2007 by Ashgate Publishing

2 Park Square, Milton Park, Abingdon, Oxon OX14 4RN
711 Third Avenue, New York, NY 10017, USA

Routledge is an imprint of the Taylor & Francis Group, an informa business

First issued in paperback 2016

British Library Cataloguing in Publication Data
Baxter, Brian, 1949-
  A Darwinian worldview : sociobiology, environmental ethics
  and the work of Edward O. Wilson
  1. Wilson, Edward O. 2. Darwin, Charles, 1809-1882 -
  Influence 3. Sociobiology 4. Environmental ethics
  I. Title
  304.5

Library of Congress Cataloging-in-Publication Data
Baxter, Brian, 1949-
  A Darwinian worldview : sociobiology, environmental ethics and the work of Edward O.
Wilson / Brian Baxter.
    p. cm.
  Includes bibliographical references and index.
  ISBN: 978-0-7546-5678-4 (hardcover)
  1. Sociobiology. 2. Social Darwinism. 3. Human evolution. 4. Environmental ethics. 5.
Darwin, Charles, 1809-1882--Influence. 6. Wilson, Edward O. I. Title.

  HM628.B38 2007
  304.5--dc22

                                                                    2006021583

ISBN: 978-0-7546-5678-4 (hbk)
ISBN: 978-1-138-25946-1 (pbk)

Contents

Acknowledgements

My thanks must go firstly to Mike Wheeler with whom I had a very valuable conversation when I was first developing the idea for this book. His encouragement and advice were extremely helpful, especially in alerting me to the developing complexities of evolutionary theory. Stephen Cowley also helped me with very useful sources of information on the sociobiological family and the developing consilience agenda as well as providing valuable comments on chapters 1 and 5. Needless to say, the responsibility for what appears in the text of this book is entirely my own.

A version of Chapter 6 appeared in 2005 in the journal *Environmental Values*, 15/1: 51–68 as 'Naturalism and Environmentalism: A Reply to Hinchman'. My thanks go to the anonymous referees whose comments on the paper helped me to clarify and strengthen the argument. A version of ideas presented in chapters 2 and 5 were read to the Third International Conference on New Directions in the Humanities at the University of Cambridge in August 2005 and were subsequently published in the *International Journal of the Humanities*, 3 (2005–2006) as 'Darwinism and the Social Sciences'. My thanks go to those participants in that conference who attended the presentation of my paper for their helpful, and encouraging, responses.

Within my own department, John MacDonald and Tony Black have supplied me with various crucial texts central to the discussion in the first part of this book, for which I am very grateful. Susan Malloch has, as ever, been very helpful in providing me with secretarial support. Paul Coulam at Ashgate has been an efficient and friendly editor, greatly facilitating the preparation of the text.

I have dedicated this book to my parents-in-law, one of whom sadly died while it was being written. I am eternally grateful to them for their love and support, and for producing their daughter, Lynn, who, as always, remains the mainstay of my life.

Chapter 1

Introduction

This book aims to explore the implications of what will be referred to as a Darwinian worldview. Charles Darwin did not intend to produce a worldview when he wrote *Origin of the Species*, aiming solely to tackle the specific intellectual problem referred to in the title of his book. Since he published that work in 1859 his scientific ideas have moved from the realm of the speculative to become received, in their twentieth-century neo-Darwinian form, as part of scientific orthodoxy.

However, many thinkers have developed from his account of the origin of species, especially as applied to our own species, a distinctive perspective on the universe that merits the label 'worldview'. A worldview embodies a specific understanding of reality, based on presuppositions that are regarded by its exponents as, at least, reasonable, and, more boldly, as firmly established or even indubitable. From these presuppositions a synoptic account is developed of the basic features of reality, encompassing the human and non-human realms, and of how they relate to each other: which are fundamental, which derivative; which are fugitive, which permanent; which have value, which do not. There are various forms of worldview, such as religions and ideologies. But not all worldviews are religious, even though they do all embody metaphysical positions, nor are they all ideologies, at least if we take the latter to involve an orientation of practical action in pursuit of socio-political purposes (Heywood 2003: p. 12). The claim that the Darwinian worldview is a new religion is one that we will have occasion to examine later in this book.

The Darwinian worldview embodies, of course, the two key ideas of Darwin's theory as applied to human beings. Firstly, it takes as axiomatic the claim that '*Homo sapiens* is an animal species'. Secondly, it accepts the Darwinian claim that this species, like all others on the planet, has arisen by a process of evolution by natural selection from an ancestor common to them all. My own interest in this worldview stems from the fact that these two claims regularly form part of the case made by environmental ethicists for the argument that human beings have moral responsibilities towards, and not simply with respect to, the non-human world of living entities. For such ethicists, in which group I include myself, we are held to be an animal species with a variety of important interconnections with the non-human world, and indeed to share a common descent with all other life-forms on the planet, as the theory of natural selection implies. This fact of our interconnection is held to justify the view that we have the obligations just mentioned (see, for example, Norton 1991). Contrariwise, we are not justified in regarding the rest of the living world as no more than a set of resources to be used for our own purposes as we see fit.

One of the main purposes behind this book is to discover whether there is any important connection between the views that share these presuppositions. Is the Darwinian worldview complementary to the environmental ethicists' project? Does it support it in any significant ways? Can reasons be found from within the Darwinian worldview for claiming both that an extensive environmental ethic is justifiable, and that we are capable of putting such an ethic into effect? Or, ominously from the point of view of environmental ethics, does the picture which emerges from the Darwinian worldview of the bases of human morality make it very unlikely that human beings are capable of putting into practice any very extensive environmental ethic (Ridley 1996)? Are we 'by nature' non-environmentalists, or even anti-environmentalists? For that matter, does the Darwinian worldview have *any* implications for morality, whether as applied to the environment or anything else?

These questions are basic to the project of this book. But, in seeking to answer them there is inevitably a whole set of other questions that have to be tackled. These are questions that are raised by the attempt to elucidate and defend the Darwinian worldview itself. To see what these are, we have to begin by giving an outline of the worldview, and this involves examining the various important conclusions which defenders of the worldview argue are to be drawn from the claims already mentioned.

The first is that the Darwinian account is capable of explaining how the appearance of design, which is so striking when the intricacies and adaptedness of life-forms is examined, may be satisfactorily explained without the need to suppose that there is a supernatural designer. Since that conception has long underpinned the most important argument of natural theology – arguing from the perceived nature of the world to the conclusion that there is a deity underlying that world – the Darwinian worldview has struck a serious blow to some important forms of religious argument.

Of course, it leaves untouched religious views based on revelation, but such revelations, as is well recognised, are problematic because they are competing. They cannot all be true, and it is highly questionable whether one can determine by rational means which, if any, actually are true. This is the central weakness of the anti-Darwinian position known as 'creationism', which seeks to claim that the account of evolution by natural selection is false, not primarily because of its inherent intellectual failings (though creationists have sought to show that it is not adequate to the facts) but because a revealed text, the Book of Genesis, says the origins of species, and of life, are to be accounted for in some completely different way.

This way involves divine creation, of course, but that is really only a detail. The important point is that creationism counters a scientific account with one based on revelation. Once this is allowed, then of course science as a whole becomes problematic, for the whole of modern scientific cosmology contains an account of the origins and age of the universe that cannot be reconciled with a literal understanding of the Book of Genesis. Yet for many people, including many religious believers, science is arguably the only form of intellectual endeavour which can produce something worthy of the term 'knowledge', even if it is not guaranteed to do so.

The recent successor to creationism, intelligent design theory, on the face of it does not face this problem, since it does not overtly claim to counter science with revelation, but seeks to remain an alternative scientific theory. It aims to show the empirical inadequacy of Darwinism, and provide empirically-testable reasons for hypothesising a designer of some sort to account for design in life-forms, in at least some instances – those supposedly manifesting 'irreducible complexity'. The latter is held to be complexity of such a kind that it cannot be derived from earlier, simpler, foundations, by blind causal processes, as Darwin's theory of evolution requires, for those bases would not have been biologically viable. Only an intelligent designer, on the analogy with human makers, can account for the existence of such phenomena, for only such a designer can effect the transition from simple materials to complex reality without the need for intermediate, simpler stages.

Those arguments appear to fail, partly because the instances of irreducible complexity upon which the theory relies are no such thing, but partly because the postulation of the existence of such designer(s) is scientifically worthless. If the designers are themselves natural phenomena, in principle accessible by normal empirical modes of observation, then we will need a scientific explanation of them in turn, and we will have in effect simply recategorised natural phenomena as artefacts. This will introduce no new principles into science. It will rather be a discovery of a purely historical nature. If the designers are supernatural, then, *ex hypothesi*, we can have no way of investigating them scientifically.

Hence, the Darwinian worldview appears to make religious belief a more problematic matter than it appeared to be prior to Darwinism. It is possible to be a Darwinian and still hold to a religious belief, but that belief cannot any longer rely for support upon the design argument. And arguably the Darwinian worldview undermines some of the other bases for traditional religious worldviews. Thus, the fact of, and experience of, morality – of a sense of right and wrong, good and bad – has often been held to justify the positing of a supernatural source for such ideas and principles. But Darwinian accounts are available, developed from Darwin's own theories, on the topic of how morality has emerged as a natural phenomenon within certain kinds of animal species. We must now, many Darwinians claim, think about morality naturalistically, not supernaturally.

In fact the implication of Darwinism, once we encompass the human species within its purview, appears to be that naturalism becomes the required approach towards the understanding of all the phenomena – intellectual, moral, aesthetic, emotional – which have been long held to be distinctive of human beings. For these phenomena are manifestations of one evolved organ – the human brain.

Whatever capacities this organ possesses must have been arrived at, if Darwinism is correct, by the process of evolution with natural selection at its core.

It is to be noted right away that it is a matter of dispute among Darwinians just how far all the biological phenomena we observe are to be accounted for exclusively by means of natural selection. But even these points do not detract from the fundamental naturalism just postulated. For, on all accounts, natural selection has been a vitally important mechanism for the evolution of species, even if not the sole one. Also, the alternative explanations proffered by Darwinians who criticise what they believe to be too heavy a reliance on natural selection are themselves naturalistic. They rely on alternative causal mechanisms – biological or cultural – which are to be located firmly within the realm of natural scientific investigation (Gould and Lewontin 1979).

Another key ingredient in the Darwinian worldview concerns the degree to which Darwinian approaches are capable of cementing a unified intellectual approach towards an understanding of human beings. It is important to note that naturalists are not co-extensive with Darwinians. Some forms of naturalism postulate important causal, and empirically accessible, processes that are nevertheless held to be disconnected from the causal processes that pervade the physical world. This is held specifically to apply to the world of human culture. This view has long underpinned the dominant idea that the social sciences and the humanities are in some fundamental way disconnected from the study of the rest of nature, even if the proponents of such views are themselves as strongly supportive of naturalism as most Darwinians appear to be (Dupré 2001). Darwinism is fine for non-human animals, but fails in the attempt to grasp human beings. 'Darwinism for non-humans, and culturalism for humans' is the slogan of this group of naturalists.

One might, then, characterise the most radical form of the Darwinian worldview as maintaining the contrary slogan, that Darwinism is appropriate for all life-forms. This is not the same as saying that we can understand human beings in exclusively biological terms. Rather, the version of the view that will command our attention is the more nuanced one that has recently been dubbed by Edward O. Wilson 'consilience'. This aims to connect up the social sciences and humanities to the biological level of understanding so as to project a unified form of knowledge at all levels. Having seen the challenge which Darwinism poses to traditional versions of religion and moral theory, we can now see the challenge it poses to received views of the appropriate ways to study human beings.

Finally, the Darwinian worldview throws up a series of challenges to our perennial search for meaning, for both our own lives and for the existence of life in general. We may call this the 'spiritual' challenge of the Darwinian worldview. In many ways it is the most important, and the most underdeveloped, of all the aspects of the Darwinian perspective. With its emphasis on contingency and chance in the account it offers of the development of life on Earth; with its emphasis on the role of competition, suffering and death in the shaping of species by natural selection; with its denial of any overall point or plan to the process of evolution, and with the blows it has struck to any reason- rather than faith-based belief in a benevolent creator, judge and saviour, Darwinism seems to many to render all life as pointless as it is painful.

Darwin spoke of the evolutionary account of life as having a certain grandeur, and that may well be true also. And if it is true, then the features just outlined which are intrinsic to it will just have to be accepted. The challenge is to find out how much of what many human beings cherish can be shown to be rationally defensible within this worldview. Truth may be there, but truth in itself can be a hard, perhaps impossible, thing to bear. Human beings want to have meaning, love and justice located in the heart of reality. Can a place be found for these within the Darwinian worldview?

In what follows an attempt will be made to investigate these issues to varying degrees. The two main issues that will be taken up will be the explanatory issue, concerning just how far the Darwinian perspective can illuminate much of importance about human life, and the moral issue concerning the Darwinian naturalist approach to the understanding of human morality. Within the latter, the question of what are the implications of the Darwinian worldview for environmental ethics will be given specific attention, for the reasons presented earlier. The issues of religion and meaning will inevitably enter into the investigation of these questions. Let us now take a further preliminary look at these two issues.

The Explanatory Issue

The first, the explanatory issue, concerns how far the phenomenon of evolution by natural selection can be used to explain human nature and human behaviour. In examining this issue the book does not consider the views of those who deny the Darwinian perspective entirely, such as creationists. Rather, it looks at the, often heated, disputes between those who are happy to accept the basic correctness of the Darwinian perspective, but who differ over how far such evolutionary history can be used to explain the nature and behaviour of human beings, as observed both currently and in the historical record.

There is less dispute, amongst those who accept Darwin's basic thesis, about the usefulness and importance of Darwin's ideas for an understanding of non-human animals and other forms of life – although some would argue that the understanding of non-human animals also requires an ineliminable non-Darwinian level of explanation. However, hackles begin to arise most vigorously when some Darwinians attempt to explain human beings in terms usefully employed in zoology and ethology.

The specific dispute which has formed the focus of much debate in this first issue area concerns the possibility of human sociobiology. In the final chapter of his 1975 book *Sociobiology: The New Synthesis*, Wilson began to apply these concepts and theories to the study of human beings, triggering a fierce debate over how far an evolutionary approach to the study of human beings is possible or useful. Philosophers, social scientists, psychologists and biologists have debated whether the human brain, and thus the human mind, is usefully conceived of as an adapted organ – that is, one containing structures selected, largely or exclusively, by the environment of our hunter-gatherer ancestors for their contribution to those ancestors' reproductive success.

By the mid-1980s there had emerged a now thriving group of researchers calling themselves 'evolutionary psychologists' who took their inspiration from Wilson and who have been engaged in the attempt to explain what they believe to be universal patterns of human behaviour on the basis of just such adapted structures. Their critics have dismissed their findings as 'Just So' stories and their general approach as misconceived, on the basis that the overwhelming influence on the development of the individual human brain is social and cultural. Hence, it is argued, the study of human beings can only properly be conducted at the level of culture and society – although there remains profound disagreement among the proponents of this general position as to how exactly such a study should be conducted.

Sociobiologists think that it is possible, and increasingly can be shown to be feasible and important, to discover structures within the human brain which in some way govern the important forms of human behaviour. These structures are held to have arisen by means of natural selection, and to have been selected, therefore, because their presence in human brains during the long period of the Pleistocene conferred upon their owners a reproductive advantage. For this explanation to work, the structures in question have to have some genetic basis. Precisely what this is in the case of each putative structure, and what the contributions of non-genetic factors may be to the development and deployment of such structures, are some of the key points under dispute.

In examining this series of debates it will be necessary to reach some view on the defensibility of what might be called the 'sociobiological' perspective on human beings. The term 'sociobiology' has fallen into disrepute in some quarters, for reasons which will become apparent as we investigate the disputes already alluded to (Midgley 2002: p. x). But it will be used in this book as a useful term to classify a developing family of Darwinian approaches to the study, and explanation, of human beings, of which the approach already noted, called evolutionary psychology, is a prominent member. As will be explained more fully later, however, this family of approaches has developed some internal diversity in recent years, so that evolutionary psychology is not the only, and perhaps not the most promising, version of the sociobiological approach. Reasons will be uncovered later for favouring the line of development favoured by Wilson himself, which has been dubbed gene-culture co-evolution theory.

An important aspect of the dispute here is over the precise character of Darwinism. The most uncompromising, although nevertheless subtle, view is that represented by Daniel Dennett (Dennett 1995). He suggests that Darwin's achievement lay in showing how the development of all living beings, including humanity, could result from an algorithmic process – a blind but orderly sifting out of actualities from the huge design space of possibilities available within the structures of physical reality.

For an algorithmic process to provide the explanation of the origins of all design features observable in actual species it is necessary to reject explanations that posit sources of design outside the processes embodied in the algorithm(s). All design features must be shown to develop from already-existing features produced algorithmically. We must, as good Darwinians, only employ 'cranes' – mechanisms resting upon, and constructed out of, earlier phases of material development, not appeal to 'sky hooks' (or a *deus ex machina*), to account for design features of organisms, including human beings. In this way, Darwinism is reductionist (all new features derive by intelligible processes from existing features) though not, Dennett argues, 'greedily reductionist' – that is, seeking to show that the proper explanation of any behaviour has to appeal solely to the first elements in the sequence of design features. A proper Darwinian explanation, therefore, of structures essential to the workings of any organism will eschew the idea of prior design. Structures emerge by the working of natural selection – an undeniably 'brute' and wasteful process involving death and destruction and, when sentience has evolved, much pain and suffering.

On this view, to be a Darwinian is to espouse algorithmic processes as one's sole explanatory tool, and to be a reductionist (though not a greedy reductionist) in one's explanations. Other Darwinians (for example, Gould and Lewontin 1979) reject the idea that one is committed to such an all-encompassing view by acceptance of Darwin's theory. For them, evolution by natural selection may explain a great deal of the observable properties of organisms, but one has to appeal to features beyond natural selection in many cases. Not all of the features present in organisms can plausibly be said to have arrived there as the result of natural selection. For example, some are side effects of other changes that were adaptations. Some have resulted from the random series of genetic changes known as 'genetic drift', the full extent of which has become apparent since we have acquired detailed knowledge of the molecular composition of the human genome (see the Appendix of this book). They claim that when one comes to the human case a whole new phenomenon – language, and the culture which that makes possible, become of vital importance in explaining human nature and behaviour (Dupré 2001).

This first issue, then, concerns a heated dispute within Darwinism. Must we reject any form of sociobiology, at least as applied to human beings, even if we are Darwinians? Or are we committed to some (subtle and sophisticated) form of sociobiology by our Darwinism? At stake in this dispute are many further issues, such as where we should stand on the idea of human perfectability if we are Darwinians, and the long-standing issue of whether biology on the one hand and social science and the humanities on the other have any important points of contact – is the latter (non-greedily) reducible to the former?

On this latter issue, Edward O Wilson has also contributed much, particularly in terms of the concept of 'consilience'. This is the idea that the social sciences and the humanities ought to be connected across organisational levels to the natural sciences, just as chemistry is connected to physics, biology to chemistry and (in Wilson's view) psychology to biology. Wilson denies that this is a 'greedy' form of reductionism, for higher levels will contain important phenomena that can only satisfactorily be explained at that level. But higher levels need to be compatible with lower levels, and lower levels may impose limits on higher ones.

In spite of the controversial nature of this view, there is often surprisingly little discussion within the ranks of social scientists themselves of the issues that it raises. For example, a recent very successful book – Marsh and Stoker's *Theory and Methods in Political Science* (2002) – which introduces the research methods of political science, encompassing behaviouralism, institutionalism, game theory, feminism and hermeneutics, contains only two references to evolution and none to sociobiology or evolutionary psychology. (Marsh and Stoker, 2002: pp. 73, 80). The plethora of approaches to the study of politics surveyed by the authors is applauded by them, and by many practising political scientists, as fruitful and unexceptionable, even though they also admit that there is often little meeting of minds, or theories, across the field.

To Wilson and his fellow exponents of consilience, such a situation rather reflects the failure of social scientists to ground their, doubtless otherwise useful, studies on a proper theoretical basis. Exponents of social science and the humanities are said to fail to grasp that their theorising has to be at least compatible with what has been discovered about human (and other animals') biology and psychology from the Darwinian perspective. Until they attempt such consilience they are doomed to continue inventing an endless series of more or less plausible theories between which choice is either arbitrary or based on the ideological preferences of particular thinkers.

In this book we will examine the current state of play with respect to this first issue area, and aim to determine how far the project of sociobiology, encompassing two of its main varieties – evolutionary psychology and gene-culture co-evolution – can be said to have met the criticisms which have been offered against it.

The Moral Issue

The second issue, the moral issue, considers what view of human morality is required – if any is – by the Darwinian perspective. This can be subdivided into two specific questions.

The first of these is how far human moral thought and action can be explained on the hypothesis that it derives from capacities that are the product of evolution by natural selection. To use Dennett's terminology, how does a Darwinian who eschews sky hooks (for example, by positing an objectively existing, perhaps divinely created, moral order which at some point begins to impinge upon human beings) explain the emergence of morality among human beings? What were the 'cranes' which were used to build it, how were they used, and what have they created? This is, in effect, the issue of how far morality may be construed naturalistically – as depending wholly upon the evolved natures of the organisms among which it appears.

This naturalistic interpretation of morality may point in the direction of moral universals, at least among human beings, if it is supposed to rest upon genetically produced structures in human brains, wherever and whenever such brains exist. Are there such universals? The answer to this question will depend upon how far it is defensible to argue, on Darwinian grounds, for a uniform basic structure to the human brain, and how far Darwinian processes of natural selection may produce variety in such basic structures.

Gene-culture coeveolution theory, as we will be discovering, allows for an interaction between culture and the basic structures of the brain which seems to allow for the possibility that different groups of humans may come to possess brains which are structured in fundamentally different ways. This might seem in turn to allow for different moral standpoints across such human groups, based naturalistically upon what those groups possess in terms of differing basic brain structures. On the other hand, it may be arguable that Darwinian approaches rule this out, and that in spite of differing cultural matrixes, human beings at all times and places share a basically similar environment which tends to produce the same naturalistic basis for the development of moral concepts and motivations.

However this question is answered, we need to note that a naturalistic approach to morality, whether Darwinian or not, faces difficulties which critics of naturalism have frequently highlighted, especially those deriving their inspiration from the Kantian tradition. Firstly, it is argued that naturalism commits the naturalistic fallacy. From the fact that human beings do in fact think some actions to be good and some bad, it does not follow that those actions really are good or bad. You cannot get from statements about what is the case by any logically valid steps to statements about what ought to be the case. Hence, moral thinking is autonomous of factual issues, whether concerning the nature of the human brain or any other phenomenon.

Then there is the troubling contingency which naturalism apparently injects into the characterisation of morality. Naturalism views morality as emerging among certain groups of animals, pre-eminently ourselves, of course. On the Darwinian version of naturalism, this is the result of blind processes of natural selection. It could have been otherwise. The existence and content of what we call morality is thus, from the naturalistic perspective, pervaded with contingency. Perhaps the human species will evolve in future to lose morality in the way that in the past our ancestors lost thick coatings of bodily hair.

From the non-naturalistic perspective, such an eventuality would represent an enormous calamity. We would have lost the key element that gives us our intrinsic value. From the Darwinian explanatory perspective, by contrast, all that can apparently be said about this is that if it produces better inclusive fitness it will in fact tend to happen. For many people this is a mode of understanding that falsifies the experienced nature of morality, which, again in the Kantian version, confronts us categorically, unqualified by any contextual references to how we in fact are placed. On this non-naturalist view, morality is encountered, not devised, however unconsciously, by moral agents.

Linked to this is another key difficulty, again bound up with the facts of human moral phenomenology. This is the experience of human freedom and autonomy. We do not view ourselves, in the course of moral deliberation, as embedded in causal networks in such a way that our moral decisions can be satisfactorily explained by reference to the causal interactions of the structures in our brains and environmental factors which impinge upon them. We experience ourselves as having the capacity for free choice, and self-determination. We can criticise ourselves for our decisions, lament our weaknesses, resolve to do better in future. Yet naturalism possesses only the language of causality, which has in it, so to speak, solely the capacity for explanation, not for justification or criticism. Thus, naturalism, it is claimed, undercuts the freedom on which alone it makes sense to hold ourselves responsible and accountable for our actions. Non-naturalism of course, having made this case, then faces its own distinctive difficulty, of providing a satisfactory account of how such freedom is possible without appeal to natural phenomena.

Such anti-naturalist arguments can be found from both religious and humanist (not mutually exclusive categories) points of view. It will thus be necessary to give full consideration to these kinds of critique, which focus upon what are perceived as the falsifying and reductionist view of morality emanating from Darwinian and other forms of naturalism. The challenge for Darwinian naturalism is to give an account of the emergence of morality among human beings which retains the contingency inherent in the evolutionary processes while coming to terms with the phenomenology of moral thought and action so dramatically highlighted by non-naturalist critics.

Darwinians in general are not amoral people. Some have argued strenuously on moral grounds for various courses of action to be pursued by individuals and groups. They cannot, therefore, readily accept any position that debunks morality as an illusion of some kind. And if the phenomenology of human moral thought is not an illusion or a confusion, and we correctly experience ourselves as on at least some occasions autonomously choosing to accept for ourselves moral imperatives which seem in some way to be categorical, then Darwinian naturalism needs some explanation of these facts.

This is not to say that Darwinian naturalism needs to accept the humanist and traditional religious interpretation of such phenomena. It may be that some debunking can go on in this field in a way that retains the centrality and intelligibility of morality to human life. But it does mean that Darwinian naturalism has to tread a narrow path between wholeheartedly accepting the humanist/religious interpretation of morality, which runs counter to the Darwinian perspective, and rejecting morality as some kind of illusion.

If Darwinian naturalism can provide a satisfactory account of the emergence and character of morality, it still has to deal with the second sub-issue, which concerns the content of human moral beliefs. Here a plethora of questions arises. Does the production of human morality by natural selection mean that it is inevitably strongly circumscribed by that origin? Is it, in some sense, always primarily to do with the well-being of the proponents of moral views? Is genuine altruism possible within it? Is it possible for human beings to accord moral recognition to the non-human? Or does human moral thought only work when human beings are its object of concern?

Questions such as these are of perfectly general significance, but in recent years they have taken on a wholly new cast. For many moralists now wish to argue for moral views which are not 'anthropocentric', or not exclusively so, in the light of what is held to be a developing environmental crisis. Some argue that, precisely because we are an animal species that has evolved by natural selection, we are both interconnected with, and interrelated to, the rest of organic life on this planet in ways that give us good reason to take the well-being and interests of such life on board in our moral deliberations. We are held to have moral obligations to life forms in general, obligations that ought to lead us to alter our patterns of behaviour in various drastic ways. In other words, taking seriously the fact of our evolution by natural selection means that we must commit ourselves to an environmental ethics, one which does not always put our own species first.

The case for this form of environmental ethics will not be fully explored in this book. I have discussed the issues raised by it fully elsewhere (Baxter 1999; 2004). Instead, attention will be focused on a key question that the Darwinian perspective raises for environmental ethics. This is whether what we may be discovering about human nature from the sociobiological approach shows that such an ethics is in some sense natural for us, or whether it rather shows that such an ethics would be difficult for us to achieve. The issue on which this turns concerns the sources of moral motivation, rather than the question of how we are to elucidate what morality requires with respect to the non-human. The two issues are logically distinct. It may be that Darwinian naturalism can provide a convincing account of the structure of moral thought – determining what counts, morally-speaking, when we deliberate over what we ought to do – and thus can make conceptual room for the emergence of environmental ethics. But it may well be that the naturalist account of morality which Darwinism furnishes will make it problematic for human beings, or most of them at any rate, to be motivated, or motivated strongly, to do what they recognise is required of them by morality with respect to the non-human.

The implications of Darwinian naturalism in this regard may depend once again on how far it licences a view of human nature as having a universal character, and how far there may be scope, via gene-culture co-evolution for example, for the development within human beings of significantly varying basic brain structures governing behaviour. If universalism is vindicated, then Darwinian naturalism may discover that there is a universal tendency among human beings to adopt an environmental ethic, in at least some circumstances. It may, of course, find no such tendency at all – though this is perhaps unlikely, given that many thinkers are devoting themselves to the development of just such an ethic – or it may find that it exists in only some individuals, or even a great many, but not universally. Another possibility is that the gene-culture co-evolution approach will find scope for permitting the development within significant numbers of human beings of tendencies to act in accordance with some form of environmental ethics. This would turn on the extent to which cultural factors can influence the development of moral motivation.

Of course, if it should turn out that Darwinian naturalism can find neither universal proclivities towards environmental ethics, nor the possibility of such proclivities developing in significant numbers of human beings as the result of gene-culture co-evolution, then the problem is not so much for Darwinian naturalism as a theoretical approach but for human beings in general. For we would be faced with the question of how we are to be brought to implement courses of action which we can recognise to be morally required but which we find ourselves reluctant to pursue. However, it would be a significant advance to discover that this was the case, for we might then have a clearer view of the precise nature of the difficulties and of how they might be overcome.

We should note that the existence of environmental ethics poses difficulties for the non-naturalist account as much as for the naturalist account. This is because a non-naturalist account, whether humanist or religious, has not hitherto been notable for any well-developed environmental ethic. Indeed, in the Kantian version it becomes highly problematic as to how any sense of moral obligation towards, as opposed to moral obligation with respect to, the non-human can be made intelligible. This is curious if morality is conceptualised as comprising categorical imperatives that are encountered rather than devised by human moral thought. The historic failure to encounter imperatives that take non-humans to be their object of moral concern requires some explanation, an explanation which may be difficult to provide from the non-naturalist perspective, given the difficulty which non-naturalism has in accounting for the origin of moral imperatives in general.

But for a Darwinian there is another important, and rather neglected, issue, namely whether the waste and suffering inherent in the process of natural selection means that we cannot properly be expected to have moral respect for a natural world within which such a process is absolutely central. Should we not seek rather to escape the nightmare of natural selection, and put as much distance as possible between ourselves and natural processes/entities? As Lisa Sideris asks, can we love a world with the often horrific workings of natural selection at its core (Sideris 2003)?

The Contribution of Edward O. Wilson

Several Darwinian thinkers have addressed the two issues – the explanatory and the moral. But often sociobiologists, evolutionary psychologists and gene-culture co-evolutionists are mainly concerned with how to generate useful and testable hypotheses concerning the evolved brain structures held to underpin human behaviour, including morality. It is hard to think of many that have had a great deal to say about the issues of environmental ethics. Contrariwise, although environmental ethicists have often focused on the fact of human beings' location within an evolved nature to seek to ground their environmental ethics, they have not usually taken a view on how far human beings, and their moral behaviour, can be explained in terms deriving from that evolutionary perspective, although one eminent environmental ethicist – Callicott – is well-known for his interest in sociobiology (Callicott 1989: p. 11).

One thinker, however, who has had much to say on both issues is someone whom we have already had occasion to mention several times – Edward O. Wilson. As a prominent biologist, he is both a proponent of human sociobiology (and has been a chief target of those Darwinians and others who wish to downplay the role of genetically-based adaptations in human life) and has extensively argued in favour of a comprehensive environmental ethic, particularly with the aim of preserving the biodiversity of the planet in the face of destructive human behaviour. His publications in the former area encompass *Sociobiology: The New Synthesis* (1975), *On Human Nature* (1978), *Genes, Mind and Culture* (with Charles Lumsden, 1981) and *Consilience: The Unity of Knowledge* (1998). His environmental ethics can be found in various publications to do with the natural world, such as *Biophilia* (1984), *The Diversity of Life* (1992) and especially in *The Future of Life* (2002).

Accordingly, in this book we will take Wilson's discussions of both issues as an object of analysis in order to ascertain whether a unified and coherent Darwinian worldview is available, and whether Wilson's is the best version of it. To anticipate the main criticisms of Wilson with respect to each of the issues, he has been accused of too reductivist a view of the relations between human biology and culture, and has been accused of too anthropocentric a view of environmental ethics. These two criticisms are obviously connected. His reductivism with respect to human moral thought is argued to rest on his view of it as heavily infected with the self-interest of the individual – kin selection and reciprocal altruism play a large role here. His anthropocentrism with respect to environmental ethics takes the form of arguing for the preservation of species and habitats for the human benefits that they promise.

He does have one distinctive form of argument that tends in a different direction, namely the case he makes for the existence within human beings of an evolved trait which he labels 'biophilia'. This still emerges on the back of human self-interest in the course of evolution, but, in Wilson's view, gives us the capacity to care for other life-forms and thus is able to ground environmental ethics on a motivational foundation of some weight. However, whether biophilia, even if it exists, is up to this task is an issue that we will have to consider carefully.

Conclusion

Clearly, there are various possibilities that might emerge from an investigation of the Darwinian worldview. With respect to the explanatory issue, it may transpire that Darwinism has nothing very useful to say about human nature and behaviour, as many of the critics of sociobiology aver. If so, the idea of consilience between the natural and social sciences would have to be rejected, and the hope of the unification of knowledge would have to be abandoned. Presumably culturalism, in all its varieties, would remain supreme in the study of humanity.

With respect to the moral issue, the failure of consilience would still permit environmental ethicists to develop their views with some reference to Darwin's theory. For it would still be possible to make some use in moral argument of the facts of the common descent across time and the interconnectedness of species at a time, which are both made use of in the articulation of the Darwinian hypothesis.

The fact of interconnectedness, of course, is not linked logically with the specific Darwinian hypothesis of evolution by natural selection, even though the science of ecology, which explores the interconnections of organisms with their environment – including other organisms – has relied upon the truth of Darwinism to illuminate its activities. Hence the interconnectedness phenomenon could still survive the intellectual demise of Darwinism itself, however unlikely that latter possibility appears at present. It is also conceivable that the claim that all life-forms share a common descent would remain unscathed even if natural selection as a mechanism is discarded after further scientific investigation, although in that case another explanation would have to be found of such a phenomenon. Hence, environmental ethics might be able to continue to appeal to phenomena currently collected under the Darwinian banner, even should that banner be incapable of entering certain areas of operation, or even should it disappear altogether.

What would be a serious blow to the project of environmental ethics would be the vindication by the Darwinian approach to moral motivation of the claim that many, or most, human beings find it very difficult to care about the non-human, at least if we set aside relations with non-human organisms that we have deliberately entered into for our personal pleasure and convenience. However, this would at least be an important finding. It would not in itself discredit the project of environmental ethics, even though 'ought' implies 'can', for we are used to the idea that some moral requirements are inevitably a strain. However, it would require us to look for some ways of lessening such a strain – including perhaps the need to accept that the best way to induce many people to act in ways which are require by environmental ethics may involve appeal to purely non-moral considerations, such as prudential ones.

This first set of possibilities involves the failure, to various degrees, of the Darwinian worldview to establish its credentials. But it may transpire instead that, with respect to the explanatory issue, Darwinism does vindicate the sociobiological perspective, to at least some degree, or in at least some areas of human behaviour. In such a case the unity of human knowledge beckons. With respect to the moral issue, the upshot of this success may still be that Darwinism will reveal that an extensive environmental ethic is beyond the reach of human beings to implement, even if they can provide a satisfactory of its rationale and content. But, more positively, it may reveal that some basis for such an ethic can be found within the evolved structures of the human brain, perhaps requiring a specific form of gene-culture co-evolution to produce it.

The aim of this book is to try to discover which of these alternative outcomes is looking the most likely at the moment. It will also try to deal with the key metaphysical problems about meaning and purpose that we have seen to lurk within the Darwinian worldview. This project is as much about mapping and delineating the issues as about reaching firm conclusions. As will doubtless become apparent, my inclination is strongly towards the acceptance of the Darwinian worldview, in spite of the unappealing aspects that many have found present within it. But it should at least be apparent at the end of this investigation that the worldview is worthy of consideration in virtue of the questions which it raises and the answers which it gives, and that the work of Edward O Wilson is an indispensable route into that worldview, even if not all his conclusions can be sustained.

Plan of the Book

In Part 1 we examine what has just been referred to as the 'explanatory issue' through a critical exploration of the basic ideas and approaches of the sociobiological family of theories. We begin in Chapter 2 with the application of sociobiology to human beings as presented by Wilson in his seminal 1975 text. This serves to introduce some of the key strengths and weaknesses of the sociobiological approach, while guarding against some of the misconceptions that surround it. In particular, we will consider, and find reasons to reject, common moral objections to the sociobiological project. In Chapter 3 we outline evolutionary psychology and use it to consider further objections to sociobiological approaches. This is done via a consideration of Dupré's critique of evolutionary psychology based on his analysis of language and culture and his championing of an alternative concept of evolution known as developmental systems theory. This critique is itself criticised, but the importance of integrating culture into the sociobiological approach to human beings is then explored further in Chapter 4 in the context of the concept of gene-culture co-evolution, developed since the 1980s by Wilson and others.

Attention is then given, in Chapter 5, to the concept of consilience, as championed by Wilson and others, which seeks to unify the natural sciences and the social sciences and humanities, via the sociobiological project. Examples are considered of what this barely developed approach might mean for the social sciences. The importance of Darwinism as a *worldview* for the successful attainment of consilience emerges in the course of this discussion. This is because of the ineradicable role of values in human life, which means that agreement on the outcome of the study of human beings presupposes agreement on the issues of values, in which rival worldviews are implicated. This topic thus marks a transition to the second issue area – that of morality.

In Part 2 we turn to the moral issue. Chapter 6 is a transitional chapter, involving a consideration of recent arguments that seek to show that Darwinian approaches to the study of human beings threaten key humanist values, and that in particular environmental ethicists have good reason to reject sociobiology, even though they wish to champion the idea that human beings are part of the natural world, not set over against it. These arguments are held to fail, but they raise the general question of the adequacy of Darwinian accounts of the origin and nature of morality, specifically of the naturalism inherent in Darwinism. In Chapter 7 we consider the pros and cons of naturalism in ethics, especially in the light of the discussion offered by Wilson of the rival Kantian approach which is so often used to indict ethical naturalists of falsifying the nature of moral experience and thought. Naturalism is defended against these attacks, and the implications of naturalism for our understanding of morality, disquieting though they are in many ways, are supported.

In Chapter 8 we examine specifically the case that can be made for environmental ethics in the light of Wilson's arguments, based as they are on conceptions, such as biophilia, drawn from his sociobiological approach. The bases that Wilson unearths in human moral motivation for the development of morally attuned environmental concern are examined, and their problematic nature is established. However, reasons are found from within the Darwinian worldview for attributing some independent moral standing to at least some elements of the natural world.

The Darwinian worldview is considered more directly in the final two chapters. Chapter 9 considers how the worldview compares with some traditional religious rivals, especially in the understanding it affords of meaningfulness in human life, and of the more specific issues of death and unmerited suffering. This leads on to the more general issue of how far the Darwinian worldview has in effect become a new religion. The limited nature of the Darwinian worldview's ability to provide consoling answers to the existential concerns characteristic of human beings is shown, and the argument is developed for the view that the Darwinian worldview is not a form of religion, but rather provides explanations of why we find the existential questions difficult, if not impossible, to answer.

Chapter 10 looks back over the discussion in the light of a parable that highlights the problematic nature of humanity's relation to the natural world, but that can also be used to show the ways in which the Darwinian worldview explains phenomena that in other systems of thought have to be taken as a brute 'given'. The contribution of Wilson to the discussion of all these issues is finally reviewed.

In the Appendix will be found a brief outline of the key conceptions of neo-Darwinism and sociobiology. Readers with little or no prior acquaintance with these fields may find it useful to read this appendix first before proceeding to the book's main chapters.

PART 1
THE EXPLANATORY ISSUE

Chapter 2

Sociobiology

As indicated in the introduction, once Edward O. Wilson and others in the 1970s began to apply sociobiology to human beings they stimulated the development of various related, but different, forms of the idea. The history and current configuration of these developments has been clearly and helpfully outlined by Laland and Brown (2002). They distinguish five members of what I will call in this book the 'sociobiological family'. These are human sociobiology, human behavioural ecology, evolutionary psychology, memetics and gene-culture co-evolution.

The relation between these is not altogether amicable. Evolutionary psychology has distanced itself from the first two on this list, charging human behavioural ecology with particularly egregious errors in its use of the idea of adaptiveness (Symons 1992). Memetics has a rather uncertain status, since it is the attempt to apply evolutionary reasoning exclusively to cultural developments, which none of the others do, even when they take account of culture, as they inevitably have to. Evolutionary psychology is the most well established and self-confident of these variations on the sociobiological theme and has attracted much of the criticism and opprobrium which human sociobiology initially received. We will examine some of these criticisms later.

In what follows we will concentrate mainly on three of these approaches – human sociobiology, evolutionary psychology and gene-culture co-evolution. These three are the most closely interrelated of the approaches. Indeed Wilson takes evolutionary psychology to be human sociobiology in all but name, although evolutionary psychologists themselves tend to disagree.

The other two – human behavioural ecology and memetics – have moved away from the paradigmatic evolutionary approach of human sociobiology. The former looks for adaptiveness in current human behaviour as it responds to a variety of environmental phenomena. The latter seeks to apply selectionism to cultural constructs directly, rather than to speculate about the forces of natural selection which might underlie, and perhaps constrain, cultural products. These two approaches may well have merits, but they each place less emphasis on the idea of a human nature which has been strongly structured by evolution, and thus are less challenging to the non-Darwinian perspectives of social scientists and others. For the purposes of this book therefore, they do not raise the key issues for discussion in as acute a way as the other three.

Let us, then, consider the main attributes of these three closely related variants of the sociobiological theme, starting with human sociobiology.

Human Sociobiology

As Laland and Brown explain, sociobiology grew fairly uncontroversially out of the field of ethology, the study of animal behaviour, pioneered by the Nobel prizewinners Lorenz, Tinbergen and von Frisch. It differed from its ethological predecessor by giving greater emphasis to the functional significance of animal behaviour – seeking to explain why evolution has selected it, and thus in what ways it may be considered an adaptation (Laland and Brown 2002: p. 69). Ethologists had been more interested in understanding the causal mechanisms that elicited the behaviour in various circumstances.

Several important evolutionary theorists have been associated with the development of sociobiology – George Williams, Robert Trivers, William Hamilton and John Maynard Smith. Richard Dawkins popularised the ideas developed by such thinkers in *The Selfish Gene* (1976). Sociobiology first applied to non-human organisms such key concepts, developed by Willams, Trivers and others, as the gene's eye view, kin selection, reciprocal altruism, optimality, game theory and evolutionarily stable strategies (Laland and Brown 2002: p. 73). George Williams had first developed the idea of the gene's eye view as an attempt to show that animal behaviour that apparently involved individuals' making sacrifices for the benefit of the group – altruism – could be more parsimoniously explained as the effect of a gene, the working of which would increase the probability that that very gene would be represented in the next generation. In other words, apparently self-sacrificial behaviour by an individual to benefit the group could really be increasing the probability that that individual would pass on its genes to a new generation.

This gene's eye view, encapsulated in the rather misleading slogan used by Dawkins, of the 'selfish gene', also underlies a whole series of other sociobiological theories and patterns of explanation. It thus underlies the explanation of altruism in terms of William Hamilton's theory of kin selection and inclusive fitness (Hamilton 1964). It is also involved in Robert Trivers's explanation of parent–offspring conflicts (Trivers 1972; 1974) and his influential account of the evolution of reciprocal altruism among small groups of animals, such as characterised the human species for much of its evolutionary history (Trivers 1971) – see the Appendix for more on this. It underlies the application by Maynard Smith and Price of game theory to the evolutionary explanation of animal behaviour in those cases in which the advantage of a piece of behaviour to the individual engaging in it depends upon what others are doing (Maynard Smith and Price 1973). From their studies emerged the important notion of the 'evolutionarily stable strategy' – the behavioural propensity which, if adopted by all the members of a population of animals, cannot be replaced by any alternative strategy.

But it was Wilson who first made a widely read proposal for the application of sociobiology to the human case in his 1975 book *Sociobiology: The New Synthesis.* The final chapter of this work, in which Wilson speculated about the evolutionary functionality of such exclusively human traits as gender differences and religion, formed part of a general case for the revamping of the traditional social sciences in such a way that they could be connected with the biologically-based synthesis represented by sociobiology. He thereby sought to bring sociobiology and the social sciences into direct contact – and perhaps conflict, given the latter's long-standing resistance to the application of biology to the understanding of human beings and their societies.

It will be useful at this point to have some idea of the content of Wilson's chapter on human sociobiology in his 1975 book, partly as a fuller introduction to the key ideas, and partly to help in the understanding of the various criticisms which quickly came to be made of it.

Wilson's Initial Account of Human Sociobiology

His opening paragraph could not have been better designed to provoke controversy, at least amongst exponents of the social sciences and humanities. For he breezily asserts that when one views humanity as a non-human zoologist might, as yet another of the various social species on this planet, the humanities and social sciences 'shrink to specialized branches of biology' (Wilson 1980: p. 271). This certainly looks to imply some form of eliminative, or 'greedy' reductionism – there is 'nothing but' biology – and that the humanities and social sciences are to be cut down to size – 'shrunk'. We will discuss what Wilson's view actually portends for these disciplines later on in the discussion of consilience, when it will become apparent that this talk of shrinking and apparent elimination is not borne out by his total view.

There then follows a description of the human species from the point of view of ecology, anatomy, and reproductive physiology/behaviour. Under each of these headings Wilson cites the unique properties of this particular species of primate, from its extraordinary geographic range to its bipedalism, hairlessness and continuous openness to copulation. But it is the extraordinary development of the human brain, and concomitant extraordinary levels and forms of intelligence that this makes possible, that Wilson goes on to emphasise. The key point here is the transformative effect of this intelligence. It is possible to find precursors of human social and intellectual traits in other primates, but the human brain has developed and altered them almost beyond recognition. Wilson emphasises the multidimensionality, subtlety, cultural formation and social complexity of human traits (Wilson 1980: p. 272).

This emphasis on the distinctive nature of human intelligence and its complex cultural dimension is an important point to recognise early on in our investigation of Wilson's account of human sociobiology. For it helps to counter any impression that may arise that Wilson believes that human beings are to be understood in the terms already produced by ethologists for the study of the other primates – in other words, that he is engaging in yet another form of eliminative reductionism, to the effect that human beings are 'nothing but' primates. His later development of gene-culture co-evolution is an attempt to do justice to this distinctiveness of human beings.

At this point he introduces the concept of 'biogram' (citing Count 1958 and Tiger and Fox 1971) to establish the purpose of human sociobiology. This biogram encompasses what he later terms the 'epigenetic rules' established in the human brain by the evolutionary process of natural selection. Such genetically produced propensities to behaviour, especially concerning the comportment of human beings with respect to the all-important environmental constraint of human society itself, have been selected because they in some way improve inclusive fitness. The genes that produce them tend to reproduce themselves more effectively than those that do not, to put the point in 'gene's-eye view' terms. Sociobiology, in other words, seeks to unearth the genetic givens of human nature that may be supposed to underlie the complexity of human behaviour in all its cultural diversity (Wilson 1980: p. 272).

With respect to this genetically produced nature, Wilson asks the questions that have become the hallmark of sociobiological investigation. How far have human beings developed genetic traits that are adapted to our contemporary world, and how far are they carry-overs from earlier stages of human development? How far have the genetic traits developed in earlier periods influenced the social constructs that we human beings have produced? How flexible are the traits in question, and in what respects? Wilson notes that the evolutionary history of hypertrophied organs in any animal is hard to reconstruct, the human brain included, and indicates that the evolutionary explanation of the bases of human behaviour will thus not be an easy matter to determine (Wilson 1980: p. 272).

The suggestion that human beings have any genetically-produced nature has in itself been a matter of raging controversy for a very long time, but once one accepts the possibility that they have such a nature and reads Wilson's program with this suggestion in mind, it is apparent that he seems to intend the questions he raises to be open ones. He also seems well aware right from the start that there are grave difficulties in the business of uncovering the correct evolutionary account of human nature. These points are worth bearing in mind when we come to encounter some of the objections to the whole family of sociobiological approaches, which often read as though sociobiologists in general, and Wilson in particular, have a naive faith in the ease with which sociobiological explanations may be generated, and a commitment to an inflexible set of genetic constraints upon human beings.

In the remainder of his discussion in the final chapter of his book Wilson undertakes to outline the general traits of the human species (presumably on the assumption that these are the most likely to have a genetic basis); to outline what was known in the late 1970s about human evolution; and finally – and perhaps very prematurely, given the confessedly speculative nature of much that follows – to consider 'some implications for the planning of future societies' (Wilson 1980: p. 272).

The first trait which Wilson goes on to examine, however, is the opposite of a rigid set of characteristics, for Wilson emphasises the immense plasticity of human social organisation, greater than anything encountered in the non-human world, and which, on his view, rests on the wide variety of human types to be found within any human society. He suggests as a hypothesis that 'genes promoting flexibility in social behaviour are strongly selected at the individual level' (Wilson 1980: p. 273). This has to be coupled, however, with the hypothesis that a wide variety of social forms can do equally good jobs of sustaining human existence. For, if they did not, then human societies would have converged upon one form. This in turn means that human societies must be able to range widely across possible forms without great harm. To account for this phenomenon he hypothesises that human beings have established such pre-eminence over other species that they have faced no strong competition for environmental resources. Hence, pretty well any reasonably stable social arrangement in which people can reproduce successfully will tend to persist. Other animals have little scope for error, but human beings have escaped narrow environmental constraints and thus have acquired room to experiment, consciously or otherwise, with different social forms.

We then receive a clear statement of the widely accepted view that the plethora of examples from history of rapid cultural change amongst human societies strongly suggests that cultural variation has little to do with genetic variation amongst human populations. The rapidity of the change is too great to be genetic in origin. Wilson accepts in large part the view of Dobzhansky that '[i]n a sense, human genes have surrendered their primacy in human evolution to an entirely new, nonbiological or superorganic agent, culture' (Dobzhansky 1963).

However, he goes on to claim that this cession of sovereignty by the genes in the human case is not total. Genetic factors have been found underlying variation in distinctive human behavioural and personality traits, such as introversion-extroversion, neuroticism, dominance, timing of major cognitive development and so forth (Wilson 1998: p. 274). If distinct human populations show even a small variation in the proportions of individuals with distinctiveness in such traits, then this may 'predispose societies towards social differences' (Wilson 1998: p. 274). This possibility requires investigation, rather than dismissal. He notes that, in the light of this possibility, it is not safe to assume that because a certain trait is absent for specific human populations, it must be purely a cultural product. In fact, its absence may originate in a genetic predisposition within the populations concerned.

A combination of genetics and anthropology is required to investigate such possibilities, which seems to be another formula for sociobiology. Since this new discipline does not yet exist, Wilson casts around for some other ways of achieving a reasonable initial sketch of the human biogram. It is important to note immediately, then, that what he goes on to offer is not meant to be part of the discipline of sociobiology. It is clearly signposted as an interim alternative. The implication is, of course, that sociobiological investigations may reveal results which will differ from those achieved by these interim, indirect, methods (Wilson 1998: p.274).

The first approach is to examine such evidence as we have already acquired concerning the 'elementary rules of human behaviour' (Wilson 1998: p. 274) with an eye to identifying the typical kinds, much as ethologists do for other animal species. He cites as an example of what he has in mind here the theory of a 'hierarchy of needs' famously postulated by the sociologist Abraham Maslow (Maslow 1972). According to this hypothesis human beings are by their nature predisposed to satisfy their needs in a certain order of priority, with most basic physical needs requiring satisfaction before the more distinctively human needs, such as self-actualisation and creativity can be attended to.

Wilson interprets this in genetic terms, with the hierarchy being determined by evolutionary forces that have selected genetic predispositions for this order of priority (Wilson 1998: p. 275). However, if this is a genuinely universal human phenomenon then it might be accounted for in non-genetic terms. Elementary human intelligence alone, it may be argued, enables everyone to see that you have got to eat, drink and sleep before anything else becomes possible. The genetic story may be true, but Wilson can fairly be charged here with a failure to consider non-genetic explanations of the phenomena he cites. This failure can be encapsulated in the idea of the 'forced move' – a ubiquitous requirement on human behaviour which stems from the logic of the common situations in which human beings find themselves, rather than from any genetic programming. The logic of the forced move explanation is an issue to which we will return later in this chapter.

The second indirect approach is phylogenetic analysis, involving the comparison of the human species with other species of primates with the aim of unearthing 'basic primate traits'. Such traits are presumed by Wilson to 'help determine the configuration of man's higher social behaviour' (Wilson 1998: p. 275). He notes the well-known pioneering efforts in this direction of Lorenz, Ardrey, Morris and Tiger and Fox, and praises them for importantly asserting the biological character of human beings. This is a fact that was ignored by dominant forms of psychological theorising of the period in which these works were written, such as that of the behaviourist tradition. But he nevertheless faults these writers for extrapolating from small, and so probably unrepresentative, samples to universal, and so faulty, conclusions.

What then should they have done? In the absence of a well-supported account of how biological traits influence social behaviour across the relevant range of closely related primate species – and, of course, this is absent in at least the human case – we need to fall back on an assessment of what traits remain constant at the level of the family or order to which *Homo sapiens* belongs, for 'these are the ones most likely to have persisted in relatively unaltered form into the evolution of *Homo*' (Wilson 1998: p. 275). When we adopt this approach, Wilson suggests, we discover that constant traits at these levels encompass such phenomena as dominance systems, with males usually dominating females, intensive and prolonged maternal care, matrilineal social organisation, and so forth (Wilson 1998: p. 275).

The important points to take from this discussion of comparative ethology are not the specific traits that Wilson cites, which are subject to on-going research and emendation (for example, recent research on the bonobo, our closest primate relative, shows females dominating males, in at least many situations (de Waal and Lanting 1997)) but the more general points he goes on to make about how such a list can properly be used when we turn to the human case. The list can be used for the framing of hypotheses about what traits have an evolved basis in human beings as well as other primates. But we cannot simply extrapolate from other primates to ourselves. Various theoretical possibilities have to be considered here.

One such possibility is that traits that are not constant between all primate species are nevertheless handed down from a common ancestor to those species that *do* happen to share them – that is, they are homologous among those species. This means that variable traits are as likely as constant ones to be genetically based. Thus, light might be thrown on the human species' evolved nature by looking at traits which are not constant across the order of primates, but are homologous between human beings and, say, chimpanzees. The converse possibility has also to be considered, namely that traits which are constant across other species of primates are not to be found in human beings.

Wilson also hypothesises that those traits which do vary across the order of primates will also be those which differ between human societies on the basis of genetic differences between human populations. That is, gene-based variability *between* species points to traits that vary *within* species too. But all of these possibilities are to be resolved on the basis of empirical investigation. Specifically, when it comes to human beings, comparative ethology can only suggest hypotheses, it cannot prove that certain traits must exist and have a mainly genetic basis in the human species – 'the comparative ethological approach does not in any way predict man's unique traits' (Wilson 1998: p. 275).

What can be extracted from these two indirect approaches to a characterisation of the human biogram is clearly very tentative and sketchy. In the light of the caveats that Wilson introduces in the course of his discussion it should be concluded that no very sure or clear picture can be given of the genetic basis of human behaviour prior to the development of proper sociobiological approaches. This point should be borne in mind constantly throughout the discussion of socobiology in this chapter. Wilson is gesturing towards a research programme, not reporting upon its well-supported findings.

Wilson then provides a fuller survey of human traits that appear to be distinctive to human beings, and speculates upon their function. In this, he draws upon the findings of anthropologists and sociologists such as Lévi-Strauss and Goffman and occasionally hints at what may be the genetic basis for the perceived pattern. For example, in the discussion of the rules for exogamy in tribal peoples, he notes that the rules allow for about 7.5 per cent of the marriages to be intertribal, which is consistent with the elementary theory of population genetics' prediction that gene flow of approximately 10 per cent each generation will suffice to counteract any tendency towards genetic differentiation of populations. It thus helps to avoid the problems involved in inbreeding. Similar outcomes are achieved in other primate groups by other mechanisms, such as random wanderings of male monkeys from one troupe to another. He does not suggest that the exogamy rules in humans have a genetic basis, but the discussion makes space for the possibility that this is so (Wilson 1998: pp. 275–7).

In other cases, Wilson is at pains to differentiate common human traits from those of other animals. For example, after cataloguing the characteristics of human heterosexual behaviour, which many are tempted to suppose is the area in which human beings are most like the other animals, and thus that there is something 'natural' about the idea that sex is for procreation, Wilson remarks that in human beings '[s]exual behaviour has been largely dissociated from the act of fertilisation'. He thus argues that those who claim that sex solely for procreation is a requirement of natural law have engaged in 'a misguided effort in comparative ethology, based on the incorrect assumption that in reproduction man is essentially like all the other animals' (Wilson 1998: p. 278).

Once again, recent research on our closest primate cousin, the bonobo, which has emerged since Wilson published his book, is revealing that this is another primate species in which sex has become divorced from exclusively reproductive activities and turned into a generalised form of social interaction between all members of the troupe (de Waal and Lanting 1997). Wilson's own point, however, is that even though much in human sexual behaviour is genetically based, it is not an automatic presumption of the sociobiological perspective that the behaviour in question is directly linked to reproductive success, which is the incautious conclusion of those who connect nature and sex-for-reproduction only. Sex may come to have a function that can be shown to improve inclusive fitness indirectly, by, for example, fostering permanent bonds of affection between male and female in a species where care of offspring is prolonged and intensive.

Wilson goes on to discuss various other traits in similar vein, covering role-playing and division of labour; communication; culture, ritual and religion; ethics; aesthetics; territoriality and tribalism (Wilson 1998: pp. 278–90). In each case we have a survey of possible related forms of behaviour in other species, especially other primates; conjectures concerning their function and evolutionary history in those species; summary of relevant research, by anthropologists and sociologists, of the variety of comparable human behaviour; differentiation of the human case from that of other species, usually involving notice of the extent of cultural elaboration and differentiation in the human case; and conjectures as to the possible genetic basis and evolutionary history of the trait in human beings.

It is important to see, as already noted, that this is not an attempt to set out the agreed findings of sociobiology on all these traits, but to sketch a very tentative outline, drawn from the findings of social science and comparative ethology, of the sort of conclusions a sociobiological approach may one day succeed in establishing. It is an account notable for its caution, emphasis on the complexity of the human case, breadth of evidential basis and boldness in its willingness to speculate about the traits discussed from an evolutionary point of view. The latter feature may have overshadowed the former characteristics of Wilson's account in the minds of his critics. Certainly, as we noted above, Wilson does not spend any time considering possible non-genetic explanations for the traits he considers. On the other hand, the whole point of the discussion is to show what a Darwinian approach to human beings might look like. The time to discuss the pros and cons of evolutionary explanations is, from this point of view, when sociobiology gets down to cases in the course of its own investigations.

Wilson offers a conspectus of current knowledge, at the time of writing, concerning the evolutionary history of human beings, a topic which, as we will be seeing, is of particularly acute concern for both sociobiologists, especially evolutionary psychologists, and their critics. As before, Wilson's account is notable for its caution. He reflects that comparison of our human ancestors with other living primates is of little direct use, as the latter live in different ecological circumstances from what appears from the evidence to have been the savannah-based lifestyle of early human beings. He argues that the best bet is to examine existing hunter-gatherer societies, as hunter-gathering appears from the evidence to have been the mode of social life of our human ancestors. We should then pick out those traits that show least variation between extant groups, on the hypothesis that those traits are the most likely to have been exhibited by our ancestors. On this basis, we have some reason to suppose that our ancestors lived in small groups organised in territories, with males dominant over females, prolonged maternal care, and a tendency to matrilineal organisation. We cannot conclude, however, that males were hunters, while females gathered.

Wilson then surveys the various suggestions from palaeontogists and paleo-anthropologists concerning the development of human traits, from bipedalism to tool-use and language, sexual division of labour, and so forth. In the course of this survey he makes the important claim that we cannot assume that human evolution has ceased with respect to either mental capacity or 'the predilection towards special social behaviours' (Wilson 1998: p. 296). A great deal of evolutionary change can take place within as few as 100 generations, or since the time of the Roman Empire. This is a point that Wilson repeats later, particularly when he comes to elaborate the theory of gene-culture co-evolution. As we will be seeing, this makes room for cultural influences to react on the human biogram, or epigenetic rules as he later calls them, to produce evolution at the genetic level. Hence, '[a]lthough we do not know how much mental evolution has actually occurred, it would be false to assume that modern civilizations have been built entirely on capital accumulated during the long haul of the Pleistocene' (Wilson 1998: p. 297). Needless to say, this also leaves open the possibility that they have been – but, as ever, one is struck by the caution Wilson shows in his speculations in this area.

In his final comments on the prospects for future society and the possible contributions that sociobiology might make to it, Wilson is rather less cautious, but he is still avowedly speculative. He sketches in the outlines of his critique of social science, a fuller version of which we consider in a later chapter on consilience. He speculates that humanity will have reached an 'ecological steady state' by the end of the twenty-first century, an idea which we will examine more thoroughly when we consider his case for an environmental ethic. He also speculates that the main contribution of sociobiology to a scientifically-informed social policy will firstly be to have delineated the evolutionary history and adaptive functions of the human neurological machinery, some of which will be obsolete, based on Pleistocene needs no longer relevant, some of which will be adaptive at the individual, but not the social levels, and some the reverse of this. We will then be in a better position to try to decide which of our behaviours are inimical to the ecological steady-state and to try to find ways of altering them. He hints that this is not to be expected to be a matter without serious difficulties – we may be faced with very difficult choices if we find that some gene-based tendencies to behave which lead us to undermine the ecological steady state are also necessary for distinctively human qualities which we have good reasons to value (Wilson 1998: p. 300). We might end up depriving ourselves of our humanity if we alter them.

This kind of speculation has come in for some fierce criticism from Mary Midgley, who finds within it a tendency to erect Darwinism into a new form of religion, or quasi-religion. We will consider this general charge in Chapter 9. But at least it shows that Wilson is fully aware that it is a difficult and complex matter to determine what exactly human traits are, and to determine what may be done to alter them, even if they do turn out to have a genetic basis.

Having outlined the general case for sociobiology which Wilson set out in his ground-breaking work of synthesis, let us now consider some of the main criticisms which have been offered of the whole enterprise.

The Critique of Sociobiology

The criticisms of human sociobiology which have most often, and most energetically, been put forward are that it is reductionist and supports genetic determinism. As Laland and Brown have argued, there is little force to these criticisms (Laland and Brown 2002: pp. 95–7). Sociobiology certainly involves the idea that genes have an important part to play in influencing human behaviour, but has always accepted that genes are not the sole influence, and that what influence they have is partly dependent upon the environment in which they operate, including the social environment. The gene's-eye view is not meant to imply, then, that genetic influences are the sole determinant of behaviour, but that evolutionary analyses should concentrate on the genes because they are the sole heritable factor. Hence, they are the sole factor upon which natural selection can be presumed to operate. It is certainly true, as we will be seeing when we consider developmental systems theory, that even this claim can be challenged, but the point is that even if genes are the sole heritable factor this in itself does not justify a claim of genetic determinism, and sociobiology does not make such a claim.

Nor is sociobiology committed to the view that higher levels of complexity can be explained entirely in terms of elements drawn from a lower level, without remainder. Rather, the claim is that the higher levels are influenced by, and may even be constrained by, the lower levels. You cannot explain all social phenomena in terms of the nature of the individuals who make up society, but the emergent phenomena at the social level may still be influenced by, and probable constrained by, the nature of individuals. Similarly, the characteristics of individuals cannot be explained solely by reference to their genes, but their genes may influence, and probably constrain, their behaviour. The key issue, then, is what these influences and constraints are, and how important they are. Sociobiology takes the view that they are probably strong and important influences and constraints. The standard social science view is that they are not (Laland and Brown 2002: p. 97).

A further confusion which needs to be cleared up in connection with sociobiology is the idea that because it claims that individuals have evolved in certain ways, and produced, as a result, certain kinds of society, sociobiology is committed to the view that what has so evolved is necessarily the best form of society for human beings to live under. But it is clear that evolution is not a process that can be characterised as inherently progressive, and the outcome of evolutionary processes can therefore meaningfully be lamented as morally insupportable, even if we in fact find it very difficult to do anything about them.

However, as all sociobiologists agree, 'evolved' does not mean 'unchangeable'. It is true that Wilson differs from some of his fellow sociobiologists in the belief that certain kinds of desirable improvement may be hard, or even impossible, to achieve, given our evolved human nature (Laland and Brown 2002: p. 99). But changeability obviously admits of degrees, and currently there is scope for legitimate differences of view as to how changeable our evolutionarily produced nature is.

The key point, however, is that, whatever view is finally arrived at on that matter, sociobiology is not committed to an endorsement of the status quo as being optimal for human beings. This, however, implies that a moral critique of the evolutionary outcome is possible, and sociobiologists therefore, need to explain how, from the sociobiological perspective, such moral debates are possible. This question raises the general issue of the tenability of naturalism in moral theory – a topic to which we will return in Chapter 7.

The other main criticisms of sociobiology are that it generates endless but untestable 'Just So' stories about the evolutionary origins of behaviour, and that sociobiologists have simply picked ad lib topics in human behaviour to speculate about before moving on to the next item which takes their fancy. The result is the failure to produce, or even attempt to produce, a worked-out body of coherent thought about human society (Laland and Brown 2002: pp. 100–106).

These are important criticisms, and sociobiology needs to explain how to test its various conjectures and explain how it is going to produce a coherent theoretical approach to the study of human society to match, say, the materialist conception of history developed by Marx. But to say that sociobiology has not done these things, or done them well enough, is not the same as saying that it cannot do them at all. It would be helpful to this enterprise if social scientists who were supportive, rather than condemnatory, of this project could seek to have an input into it. Those whose training is primarily in biology may simply not have enough familiarity with the social science literature and perspectives to proceed much beyond the piecemeal approach already noted to be inadequate – or at least, to be inadequate as a final stage.

Social Science's Hostility to Darwinism

However, this whole Darwinian approach has to face some formidable opponents waiting for it, especially in the form of culturalist 'bouncers' whose job is to eject any natural scientists with the temerity to intrude upon the space reserved solely for social science. Most social scientists are non-professional Darwinians – they happily accept the Darwinian approach to the study of all other life-forms, and, within narrow limits, to its application to human beings. But typically they do not accept that biology has any profound implications for human psychology, and so in particular they do not accept that human beings have brains with various forms of prepared learning and specific capacities hard-wired into them. Sociobiological approaches, such as evolutionary psychology, follow William James in postulating that the human brain contains many more instincts that do other animals, not fewer (Tooby and Cosmides 1992, p. 93). Social scientists have come to believe that human beings have rather fewer instincts than those of other animals, and those which they do have are not of great importance in understanding human life.

On the mainstream social science view, the crucial element which makes the difference between human beings and other animals is the human possession of language. Language allows the creation of culture, embodying forms of knowledge and traditions that pass down through generations of human beings. It permits human beings to acquire theoretical thinking and thus the propounding and testing of general explanations of how their world works which in turn enable them to break away progressively from the environmental constraints within which they would otherwise be confined.

Humans change their environment, thus changing what they are supposed by Darwinism to be adapting to. They have done this ever since they have been human, and thus their evolutionary history cannot properly be understood as the acquisition of a multitude of specialist adaptations. The human brain is thus produced, in the case of each phenotype, by the developmental interaction between an all-purpose language generating device and the specific, culturally created environment in which it comes to maturity.

The resulting preferred view of humanity is a strongly culturalist one, embodying the Durkheimian claim that social phenomena can only be properly explained at the social level, not the level of individual psychology, and the Wittgensteinian view that language is an essentially social phenomenon – there cannot be a logically private language, all meanings are public, specified by interpersonal, and changing, criteria embodied in language games which make up a distinct, and often unique, form of life. To understand society is to grasp the system of interpersonal rules that structure the games played within it (Durkheim 1982; Wittgenstein 1953; Dupré 2001: pp. 33–7). We will defer consideration of this kind of argument until the next chapter, when we will consider a version of it championed by John Dupré in his criticism of the aims of evolutionary psychology.

However, although many social scientists, of all stripes, would seek to put forward theoretical objections to sociobiology, or, indeed to any other theory which purported to deploy Darwinian concepts, such as adaptation and natural selection, in a project to develop extensive explanations of human beings and their behaviour, there is also a strong tendency to find such Darwinian approaches to be morally and politically objectionable. That is, the Darwinian approach is held to be objectionable for what are thought to be its normative and prescriptive commitments. Let us turn to this issue.

The pattern of argument here is very familiar. Darwinism's emphasis on competition between organisms and the pursuit of their self-interest, within the course of natural selection, coupled with what is thought to be a gene-based and determinist analysis of the structures within the human brain, seem to lead inexorably to the view that there is an unavoidable, genetically-determined tendency to selfishness among human beings. This, when coupled with the Darwinian view that some organisms will be better equipped to compete than their rivals, seems to provide a solid biological basis for the existence of unavoidable human inequalities which explain differential outcomes in attainment of life-chances. Taken together, these empirical claims about what human beings are like seem to justify the normative claim that a form of social organisation that is highly competitive and permissive of the pursuit of self-interest is the best for human beings – because it is the most natural for them – and thus seem to justify the prescriptive claim that human beings ought to organise their societies so as to realise this value.

Many social scientists hold very different normative and prescriptive views from these, and thus, if they accept the version of Darwinism's implications outlined in the previous paragraph, they will be inclined to resist very strongly the claims of Darwinism to have an accurate grasp of human nature. Some who have been influenced by Darwin's theory – the proponents of Social Darwinism – do seem to have put forward a version of Darwinism for human beings of the kind just mentioned, and to have drawn the normative and prescriptive conclusions which many find so objectionable. But many other Darwinians have argued just as strongly that natural selection does not restrict organisms to selfish and competitive behaviour, and that even if it did, the normative and prescriptive conclusions just mentioned do not follow. Let us review some of these ripostes.

The first port of call is the naturalistic fallacy. Even if we can show beyond all reasonable doubt that human beings have ineradicable tendencies to competitive and selfish behaviour this does not show that we should tolerate or encourage such tendencies. Human beings have a tendency to commit rape and murder – at least, these are perennial and pervasive aspects of human behaviour, however we explain them – but we do not suppose that because these are 'normal' for human beings that we must therefore permit or celebrate them. We cannot get from 'is' to 'ought' by logic alone.

But, in any case, it can be demonstrated that natural selection does not always select competitive, selfish behaviour. We cannot argue a priori that, in any given environment, inclusive fitness will always only be shown by the organism which has the most competitive, selfish, tendencies. Organisms, including human beings, plainly do act to benefit others, even at cost to themselves, and show caring and loving forms of behaviour. This behaviour, if an adaptation, will have been selected on the basis of its genetic origins. Hence, it might be said to be 'really' an example of the selfish gene in operation.

It was noted earlier that the phrase, 'selfish gene', championed by Richard Dawkins is misleading. That is because genes are never really selfish – for one thing, they aren't even selves, but sequences of DNA. They have no self-awareness, intention or cunning. They just are, and the only question a Darwinian needs to consider concerning them is how well they are going to be able to reproduce themselves in a given context. Successful reproducers are called 'selfish' as a form of shorthand, metaphorical, anthropomorphism.

What 'selfish' genes produce in the phenotype, however, is often *really* altruistic and loving behaviour. Altruistic organisms really do put the interests of others ahead of their own. In the case of human beings, they often do this consciously and intentionally. The challenge for Darwinism has been to show how altruism, love and care can be produced by natural selection. What has to be shown is how genes 'for' that behaviour can be expected to replicate more successfully than others do in certain environmental contexts. As we have already noted at the start of this chapter, the theories of kin selection and reciprocal altruism show how this is possible (Dennett 1995, pp. 477–81), and show how the human moral sense may develop, with its panoply of traits such as guilt, shame, righteous indignation and so on.

Yet, to repeat the first riposte, this does not licence a normative or prescriptive conclusion to the effect that it is good, because natural, to be only altruistic and caring to intimates, ignoring strangers. It does mean that if we wish to encourage altruism beyond its natural scope ('natural' in terms of Darwinian explanation) we will need to be alert to the problems our evolved natures pose for us. To put the point in terms of a useful distinction of Richards, we may need to alter our derived principles (for example, concerning how children are to raised/educated, how societies are to be organised more generally, and so on) in the light of our Darwinian self-understanding, in order the better to achieve our fundamental principles (social justice, say, and various forms of liberty and equality) (Richards 2000: pp. 239–40).

However, it is necessary also to note that what contemporary accounts of gene-based structures in the human brain lead to is not a view of such structures as rigidly fixed or incapable of alteration. Indeed, as Ridley has recently shown, the distinction between what is reversible/alterable and what is fixed does not coincide with what is largely genetic (with low environmental input) versus what is largely environmental (with low genetic input). Some developmental phenomena in the human brain may depend very largely upon environmental, including cultural, inputs for their specific character, and yet be very hard to alter/reverse once this character has been achieved. Some that are largely genetic in origin (although with somatic and/or external environmental inputs) are relatively easy to block, alter or reverse (Ridley 2003: pp. 144–9).

Thus, even if sociobiology does discover elements in the brain that are largely genetic in origin, this in itself may not show anything about how easy it is to block, alter or reverse the development of such elements, if we discover good reason to believe that their existence constitutes a serious impediment to the attainment of some moral aim.

In addition, it is increasingly becoming apparent that the successful explanation of how human beings acquire culture throughout their lives involves tracing the development within the brain of neural structures. These develop by means of the operation of genes within the neural cells as they respond to stimuli from the environment, both within the cell and from outside the organism entirely (Ridley 2003: pp. 133–44). That is, the genes do not simply create the phenotype and then cease activity, but are constantly operating within the cellular environment throughout the life of the phenotype. Ridley refers to this as 'nature via nurture' (Ridley 2003: p. 4). Thus, even if we restrict ourselves to some form of culturalism, we are now faced with the claim that genes play a crucial role in the acquisition of culture. It is at least prima facie possible that their indispensable involvement may have some explanatory role to play in the forms of culture which human beings generate. An evolutionary dimension is built into such a perspective by the theories of gene-culture evolution that we will consider in Chapter 4.

Many fear that gene-based forms of explanation commit us to the view that certain human behaviour patterns are unalterable. But defenders of a Darwinian perspective can point to the strong possibility that there are what Dennett has usefully labelled 'forced moves' which constrain human life in various ways, and which have nothing to do with genes (Dennett 1995: p. 226). A forced move is an action that is unavoidable if certain aims are to be realised (such as avoiding checkmate in certain circumstances in the game of chess). Although the idea of a forced move does not address the fears about genetic fixity directly, it does at least indicate that important kinds of fixity may not have anything very much to do with the Darwinian perspective at all. Forced moves, where they exist, will make all human groups in all times and places show similar patterns of behaviour, at least as long as they retain characteristic human psychological traits.

Perhaps human social life is permeated by situations in which forced moves are operative. For example, perhaps the fact that some form of market economy is all-pervasive in human life has nothing to do with the existence of a brain module 'for' market behaviour that developed as an adaptation, but is simply a reflection of the fact that markets are the only workable means available to human beings for engaging in economic transactions in an efficient and flexible way. As long as human beings seek such efficiency and flexibility, this is the only way they can do it. Hayek, among others, offers various arguments to this effect (Hayek 1949: pp. 77–8).

Clearly, if this is correct, then the existence of such forced moves, given 'normal' human values and purposes, may mean that certain political projects aiming to realise certain fundamental values, face serious difficulty. The example of markets shows how the logic of the 'forced moves' idea may be used in the service of certain conclusions commonly found on the right of politics (although not exclusively there, as the case for market socialism shows – see Miller (1989)). However, such arguments are also frequently deployed in the service of political projects drawn from the left of politics. For example, the argument that only an economic system in which the means of production are collectively owned can be non-exploitative is essentially a 'forced move' argument.

As we noted above, Wilson in his exposition of the case for sociobiology does not consider the idea of forced moves as an alternative explanation for universal patterns of human behaviour. This is a point often made by critics of Wilson and of sociobiological approaches, for the existence of the 'forced move' phenomenon obviously makes possible an alternative explanation of the existence of universal traits from that which rests on the existence of genetically-produced hard-wiring in the brain. This, however, does not detract from the main point here, which is that those suspicious of the idea that Darwinism seeks to foist upon us a picture of human possibilities that are constrained by what we are forced to do by our genes must also recognise that such a threat can come from other sources, and that their own theoretical positions often require at least some such arguments. It may also be that a 'forced move' argument is in any case relying at one remove on the idea that there is something fixed in human nature which provides the context of values and purposes relative to which the 'forced move' becomes forced in the first place. In that case, the sociobiological style of explanation may actually be required at a more fundamental level to make the forced move argument work.

We can conclude this discussion of some of the chief moral objections to the application of Darwinian ideas to human life by saying that there are various possible sources of at least relative 'fixity', and hence universality, in human life. Some may be primarily gene-based, perhaps operating via the creation of modules in the brain of the kind posited by evolutionary psychology. Some may be primarily environment- and culture-based, albeit still requiring the operation of genes to be effective. Some may be forced moves, inherent in the logic of the situation, or some other extra-mental factor.

As Wilson concluded in his chapter on human sociobiology, knowing what the (relatively) fixed points are, and what fixes them, is plainly an indispensable prerequisite to coping with them – by reducing their effect, circumventing them or attacking them directly if what is fixed is also deemed to be an obstacle to the fulfilment of desirable goals. We may seek to alter the effects of genes by, for example, genetic engineering, or by pharmacological means. We may seek to alter the cultural or other environmental factors that, in interaction with genes, produce fixed patterns of behaviour. In the case of forced moves, we may seek to alter the contextual ingredients in human nature relative to which the fixity obtains and/or probe the situation for elements of elasticity which allow us to adopt new, and more desirable, behaviour patterns.

The point is that in all this, the Darwinian approach does not threaten to throw elements of fixity into what is otherwise a perfectly plastic situation. Even if Darwinism had never been thought of, and we knew nothing about genes, we would still be struggling to discern, and to argue over how best to cope with, elements of fixity deriving from cultural and environmental factors operating upon human beings and from the putative forced moves that they face. What an acceptance of the Darwinian perspective opens up is a more empirically adequate grasp of the causes of relative fixity in human nature, and thus by extension of the forced moves we face.

Social scientists, therefore, should not reject Darwinist approaches to human social and political phenomena on moral grounds, for there are no cogent moral objections to this approach. Those that are offered are based on false pictures of both Darwinism and our actual situation. However, we need still to consider the objections to sociobiology based on the claims concerning the cultural basis of human behaviour. The next four chapters will largely be taken up with exploring that issue in all its facets. Some of the issues can profitably be raised in the discussion of evolutionary psychology, to which we turn next.

Chapter 3

Evolutionary Psychology

As Laland and Brown explain (2002: p. 154), in the 1980s some theorists who were attracted by the possibility of applying evolutionary thinking to the study of human beings began to differentiate their approach from that which they associated with human sociobiology of Edward O. Wilson. Wilson's error was perceived to be that of seeking to conceptualise human behaviour as directly the result of natural selection, rather than as mediated by psychological mechanisms that were held to be the real objects of selection.

As this group was dominated by academic psychologists the name 'evolutionary psychology' was devised to label the approach. Even if Wilson's views were regarded as representing a false route, the ideas of other major contributors to sociobiology – Trivers, Hamilton, Williams – were regarded as important to the new discipline. Clearly, in making the attempt to apply the ideas of such thinkers to the human case evolutionary psychologists were at least following in the spirit of the Wilsonian enterprise.

The psychologist Leda Cosmides and the anthropologist John Tooby have been associated with evolutionary psychology from the start. The views of their discipline which they have propounded contain some striking themes, the most prominent of which concern our evolutionary history. The human brain, they argue, has evolved under the influence of natural selection, and thus its characteristics are best understood as representing a series of adaptations to the environment in which human evolution took place – the EEA: environment of evolutionary adaptiveness (a term taken from Bowlby 1969). They argue that the period represented by recorded human history has been too recent to have had any extensive effect on brain evolution, as compared with the millennia-long period of our hunter-gatherer ancestry. Hence, there is no good reason to expect that our ancient brains will be well adapted to the rapidly changing recent environment. Recognisable adaptations to our contemporary world may be hard to find.

To work out what the adaptations were which were evoked in our ancestors by the EEA during the last two million years we need to determine what kinds of problem our ancestors would have faced, and then what adaptations would have solved those problems. Working out what these adaptations are is a matter of determining what psychological mechanisms would be necessary to solve the problems identified from the EEA. Hence, the project of evolutionary psychology is to hypothesise the existence of such mechanisms and to locate them in the actual human brain. This, if it works, will then give us a good understanding of how the human brain actually functions, and thus the key to understanding human behaviour, including, of course human social behaviour.

From the then developing theories of cognitive psychology, that found it illuminating to consider the human brain as an information processing device, evolutionary psychology took another of its central themes, which is to view the mental mechanisms that it hypothesises to exist as containing a great deal of prepared learning, or embodied information. Cognitive psychologists and investigators into the creation of artificial intelligence had discovered how much a cognitive device had to contain in the way of prior information in order to engage in even the simplest of cognitive functions (Laland and Brown 2002: p. 155). Cosmides and Tooby emphasised that this implied that the human brain was not a blank sheet at birth, waiting to be written on by experience, but that it came into existence with psychological mechanisms which already contained a good deal of information. If we use the term 'instinct' to refer to the preparedness of a brain to act on the basis of information hard-wired into it, then we can say, with Cosmides and Tooby, that the human brain contains more, rather than fewer, instincts than do the brains of other animals (Tooby and Cosmides 1992: p. 113)

Another key implication of this, in turn, is that there are universal features of human brains, or a common human nature, which underlies and produces the diversity of cultural manifestations which we see around us and learn about from the records accumulated by anthropologists and others. Several attempts have now been made to at least sketch the outline of this universal human nature, such as the works of Brown (1991), Wright (1995) and Pinker (1997).

The modular theory of the mind characteristic of evolutionary psychology posits the existence within the human brain of a large number of mechanisms for processing various forms of information. Given that these are held to be adaptations to the EEA, it can be expected that at least some of these will have some clear reference to the conditions presumed to have held during that period. Hence we find the claim that fear of spiders and snakes, even among people living in parts of the world with no poisonous spiders or snakes, derives from the operations of an in-built mechanism which would have served our ancestors well in the parts of the world where they did evolve, and where such creatures were common. Preferences for living in certain kinds of landscape, familiar to our African ancestors, of natural parkland, with some high ground and bodies of water, can be attributed to an adaptive mechanism which would have led us to gravitate towards landscapes within which we had the best chance of surviving, and thus which it would be beneficial for us to derive pleasure from living in.

Both of these examples have fed into Edward Wilson's concept of biophilia – of heightened sensitivity, of both a negative and positive kind, to living forms around us. Other mechanisms that have become specifically associated with evolutionary psychologists are strongly directed towards our fellow humans. These comprise such phenomena as our sensitivity towards the tendency of others to cheat us, the various desiderata which males and females look for in their sexual partners, and the capacity to learn any human language to which we are exposed from infancy.

Such mechanisms are supposed to be at least relatively discrete from, even if interconnected with, others, and to be relatively unchanging and relatively hard to change. The first of these points is another matter of controversy, for the 'grain' problem (Sterelny and Griffiths 1999: pp. 322–3) plagues it. This implies that there is no undeniably single way to delineate the mental phenomena that are to be assigned to a putative mechanism or module. 'Aggression' for example occurs in many mental states and can be seen to have a variety of functions, or perhaps none. What is true of such a trait is arguably a common feature of all mental states, and that suggests that the decision to investigate one rather than another aspect of a complex mental phenomenon, which we have to do to apply the modular theory, may well be entirely arbitrary. Other researchers might decide to divide up the human mind into an entirely different set of components.

In reply to this point evolutionary psychologists have available to them an argument, articulated by Laland and Brown, which we will find to be of service later in connection with the problem of trying to divide culture into discrete entities. The argument is that this is how science has always had to proceed, and that it is not a hopeless prospect to attempt to subdivide any holistic phenomenon, even though mistakes may be made, and interconnections which make things complicated must always be expected. In any case, the study of human universals shows that there is strong prima facie evidence that the division of the mind into different components by 'folk psychology' – the version of psychology that is embedded in common language and thought – shows great similarities around the world. This suggests that it is tracking a set of real distinctions, not completely arbitrary ones. However, difficult it may be to operationalise such distinctions for theoretical purposes, they do not represent an inherently flawed starting point.

The question of how difficult it is to change mechanisms is part of the more general issue of genetic determinism. As we saw in the last chapter, 'difficult to change' does not mean 'impossible to change', and the fact that a mechanism developed naturally in certain conditions within the normal course of human development implies nothing in itself about how easy it is to block, change or divert that development within the life of a particular individual. However, it is reasonable to expect that, if a mental mechanism is an adaptation, and thus has been of benefit to individuals of that species in the past, it will not easily be changed or diverted, on pain of losing its adaptive function.

As well as claiming that there are such mechanisms in the human mind, it is inherent in the logic of the evolutionary psychologists' position that such mechanisms contain a great deal in the way of prepared learning. To put the point in another mode, the human mind is prepared by its evolutionary history to attend to and deal effectively with some things rather than others. In the well-known metaphor of the Swiss Army knife, the human mind is regarded as being made up of a variety of specialised components, pre-designed to enable individuals to deal effectively with only some of the specific problems with which it had to cope in the EEA.

As Laland and Brown show, there is a lot of evidence that non-human animals do indeed manifest this prepared learning (Laland and Brown 2002: p. 163). There is some good evidence, particularly in the field of language acquisition, that human beings also manifest prepared learning. At least, this is how evolutionary psvchologists have treated the Chomskyan idea that all normal human beings posses a language-acquisition device which enables them to absorb and master their native language by about the age of five, even given the poverty of the linguistic data to which they may be subjected. That such a fantastically complicated feat as learning a language from nothing can happen so readily does indeed provide overwhelming evidence for the existence within each brain of a mental module containing a great deal of prepared learning.

The controversies start to arise when this claim is accorded completely general validity. How many other specialised mental modules there are, each containing prepared learning, is a matter of considerable dispute. The alternative hypothesis, characterised by evolutionary psychologists as the 'standard social science model', is that the human brain is largely a general purpose learning device (Tooby and Cosmides 1992: pp. 24–31). This minimises the prepared learning needed and gives the brain an in-built flexibility. But the question is whether such a device could actually do all that is necessary to enable its owner to survive in the real world.

It will also be a pertinent question, at least to be pressed against those who accept the general claim that human beings evolved through natural selection, as to how such a device could have evolved at all, especially as the realm of the non-human from which we are believed to have evolved is, as already noted, permeated with examples of prepared learning. It is perhaps more plausible to suppose that natural selection would work as usual by adapting what is pre-existing, and if this was domain-specific mental modules, then what would be expected is more and more elaborate modules of this kind, rather than a new phenomenon entirely. But, of course, this too is a speculation, and we will be noting shortly that there are some strong arguments on the other side of the issue.

What of the concept of the environment of evolutionary adaptedness? To a large extent, within evolutionary psychology it represents the place to look for clues as to what adaptations would have been needed for survival by our ancestors, and thus for clues as to the nature of the mental mechanisms with which humanity is currently endowed, maladapted as they may well be to our contemporary lives. However, the problem with treating the EEA in this way is absolutely obvious – the EEA has gone forever, and its actual nature is largely, and always will be, a matter for speculation. Such speculation need not be either stupid or ill-informed, of course. The course of history prior to human records is an area where a whole host of scientific disciplines have found it possible to contribute to the generation of a complex overall picture of the processes at work through time and the state of various parts of the planet at a time. Although human records were not kept, and for most of it human beings were entirely absent, still the processes leave a trace and the fruitful way to interpret the traces can be worked out with sufficient patience and ingenuity.

Hence, the idea of using the EEA as a locus for ideas about environmental problems and possible solutions in the form of adaptations is not a complete non-starter. But it does not lend itself to easy speculation. Those who accuse all members of the sociobiology family of peddling 'Just So' stories have a point, even if one might well argue that they tend to overplay it. This is an issue to which we will return shortly when we consider the criticisms offered by John Dupré to the claims of evolutionary psychology. In particular, as Laland and Brown note, we are very ignorant of the precise mode of life of our human ancestors, although we do know, for their remains have been found, that they inhabited a wide range of habitats (Laland and Brown 2002: p. 178). However, whether they were actually for any significant periods of time hunter-gatherers in the sense which has been applied to modern peoples in the anthropological literature is a matter of serious dispute (Laland and Brown 2002: p. 178). Ignorance of their mode of living makes speculation about the environmental pressures that they faced, and thus of what kinds of adaptations they may have evolved, very difficult, if not impossible.

Laland and Brown point out that Cosmides and Tooby claim that the EEA should be viewed as a composite of the various habitats our human ancestors occupied during the Pleistocene (Tooby and Cosmides 1990). In reply to this, Laland and Brown make the valid point that to the extent that we conceive the EEA in this way we lose any real capacity to use it to cast light on the environmental pressures that our ancestors faced. They also argue that it is reasonable to speculate that many characteristic human traits may have a long evolutionary history, and be shared with many of the other species with which we have a recent and not so recent common descent. Hence, once again, it will be tricky matter to show that any given mechanism must have evolved as a response to some feature of our human stone-age ancestry (Laland and Brown 2002: p. 179).

They report that, in reply to such points, John Tooby has argued that the EEA concept needs only knowledge of the Pleistocene period in order to reconstruct the history of the relevant human adaptations, since any pre-Pleistocene trait would not have remained constant prior to the Pleistocene, but would have been evolving. As Laland and Brown point out, this view involves such assumptions as the following: that pre-Pleistocene traits were not unchanged; that no important human mental mechanism embodies an inheritance predating the Pleistocene; that no important evolution in such traits has taken place since the Pleistocene and that there are no variations in the rate of evolution (Laland and Brown 2002: pp. 180–81). The status of such assumptions is not easy to gauge, but they are, as Laland and Brown note, clearly controversial.

This point is given force by Wilson's attempt in the final chapter of *Sociobiology: The New Synthesis,* to extrapolate backwards from the shared traits of contemporary hunter-gatherers to claims about the traits of our distant hunter-gatherer ancestors. For the few features he cites by this route – male dominance over females, prolonged maternal care, pronounced socialisation of the young, matrilineal organisation – strongly overlap with the ones he postulates for those ancestors on the basis of the alternative route of comparative ethology, which reveals traits shared with other primates (Wilson 1980: pp. 276, 282). This suggests that those traits have their evolutionary origin at a point before human beings became a distinct species.

Setting aside these problems which face the EEA concept, we need to consider how evolutionary psychologists seek to put their ideas to work. As Laland and Brown point out (2002: p. 165), Buss (1999) has distinguished two techniques. The first is to begin with well-established theories in evolutionary theory, such as the kin-selectionism of William Hamilton, and derive from them, in conjunction with well-supported hypotheses about relevant aspects of humans' ancient environment (such as their habitation of small groups) predictions about what mental mechanisms they would need to engage in such kin selection. Then it will be necessary to embark upon empirical testing to see whether in fact such mechanisms exist in actual humans, of the kind predicted by the theory.

The alternative is to begin with an observed pattern of behaviour and to generate a hypothesis concerning its evolutionary origins which will then lead to the formulation of testable predictions. This is a matter of taking something as given and speculating about the forces which brought it about – what Pinker refers to as 'reverse engineering' (Pinker 1997: pp. 21–3). It is what the armed forces of a state might do when they capture a previously unknown war machine of an enemy – namely try to discover how it was made. In each case, however, the theories and hypotheses need to be tested against actual human beings, in order to see if the putative mechanisms exist at all, and whether they work as they are predicted to. In this respect at least, whether 'Just So' stories or not, the speculations of evolutionary psychology aim to meet the normal requirements of empirical science.

Nevertheless, there are further difficulties of principle which make the application of these techniques to human beings rather problematic. We have already noted the difficulties that stem from the paucity of knowledge we have, or can have, about our ancestors' lives in the Pleistocene. But the interaction between culture and genes emphasised, as we will be observing in the next chapter, by the gene-culture co-evolutionists means that it is impossible even in principle to take the non-human environment as a given factor against which it is safe to speculate about the kinds of problems with which our ancestors would have had to deal. We have for long inhabited a wide variety of natural environments and have for just as long been engaged in actively modifying them. This means, as Laland and Brown note (2002: p. 182) that we must have evolved an unusual degree of flexibility and adaptability which would in turn have enabled us to adapt our environments in new ways, thus altering the selective pressures upon ourselves.

This leads on to the more general claim that the evolutionary psychological approach to understanding the human brain has become too committed to the idea of modules which are highly specific with respect to behavioural domains.

As we have seen, there are good reasons in some cases to suppose that there are such domain-specific modules, such as the language-acquisition device. But, as Laland and Brown argue, sometimes the efficient solution to the problem of designing a mechanism to enable human beings to interact successfully with environmental elements will be to produce a domain-general mechanism (Laland and Brown 2002: p. 183). As they correctly note, 'specific' and 'general' are not polar opposites, but end points of a spectrum. In this respect, the disagreements between evolutionary psychologists and their opponents are to do with differences of judgement about precisely where on this spectrum it is defensible to place a putative mental function. This is, however, a different dispute from that concerning whether there are any, or many, modules at all, and it is important not to confuse them.

The idea of a learning capacity is obviously crucial here. Even though such a capacity is clearly one which has evolved in the human species, it is, as Laland and Brown argue, unlike other mental capacities precisely in the respect that its function is to seek out and retain information relevant to efficient interaction with the world, information which could not be prestored in the brain by some genetically-controlled process (Laland and Brown 2002: p. 184). It is clear why it is advantageous to an organism to have such a capacity. It is still possible to specify the workings of such domain-general learning systems by means of suitably generalised epigenetic rules, such as the example cited by Laland and Brown, Edward Thorndike's 'Law of Effect': 'Actions that are followed by a positive outcome are likely to be repeated, while those followed by a negative outcome will be eliminated' (Laland and Brown 2002: p. 184). However, once the rules embedded in the system reach this degree of generality, the explanation of specific instances of human behaviour will largely turn on the specific information garnered by the information-acquisition system, information which the individual has acquired and not simply activated.

This then creates an additional problem, for now behavioural traits will be found to depend in some way upon brain structures, and yet it may be wholly unclear whether those structures are to be explained as hard-wired domain-specific modules or as acquired characteristics assembled by some general purpose module or other. The ubiquity of the trait among human beings will not necessarily be a guide as to which it is, for a general purpose module may often be expected to produce the same results in individuals as the result of its encounter with a sufficiently shared element of the human condition.

On the face of it the dispute over how far the human brain may be organised along domain-specific versus domain general lines is a purely intellectual one. Further, the difference between evolutionary psychologists and their critics on this issue may seem to be one purely of degree, especially given the comment from Tooby and Cosmides, reported by Laland and Brown, to the effect that many psychological traits are indeed domain-general, and that their own work has revealed just that (Laland and Brown 2002: p. 186). But many will regard the issue as one which has profound implications for how much of human behaviour may be instilled by purely cultural processes, and how much is hard-wired into the brain, and thus, it is supposed, unalterable. This is where the issues are taken up into wider ideological positions. Suffice it to say at this point that this is a dispute within the Darwinian worldview. Neither side disputes that the human brain is an evolved organ. We will return at later stage to discuss this issue, a version of the time-hallowed nature–nurture dispute.

A further important series of objections to evolutionary psychology, at least as it has been practised up until now, is that it is working with an outmoded model of what is involved in the biological process of evolution. As Laland and Brown explain, evolution can no longer be supposed solely to involve the production of adaptations by natural selection, where an adaptation is any trait that increases inclusive fitness.

Many traits which organisms exhibit are not adaptations. Some are side-effects of adaptations, or exaptations; some are the result of random genetic changes, or 'genetic drift' (Laland and Brown 2002: pp. 187, 193) (for more on this see the appendix). Further complications stem from the discovery of a variety of levels at which natural selection operates. Laland and Brown cite Endler's claim that there are at least 21 processes involved in natural selection, some below the level of the individual organism and some above it – at the levels even of the species and the clade (Laland and Brown 2002: pp. 187–8). Further, the key concept of inclusive fitness, the definition of fitness devised by William Hamilton which is usually favoured by evolutionary psychologists, turns out to be only of limited use and other models have had to be devised to deal with other problems (Laland and Brown 2002: p. 187).

Clearly, if it could be assumed that all observable traits of organisms are adaptations in the Hamiltonian sense it would still be problematic to work out what selection pressures could have produced them, but it would be safe to assume that there must be some such explanation. Now that this presupposition turns out to be untenable, it cannot just be assumed that any observable trait is an adaptation, and thus the difficulty of devising testable evolutionary hypotheses becomes even more fraught. Laland and Brown suggest that a variety of forms of evidence should be looked for in support of any adaptation hypothesis. The use of mathematical models (the gene-culture co-evolutionists' strongpoint, as we will be seeing in the next chapter), the comparative method, phenotypic manipulations, reverse-engineering hypotheses, and so on are all recommended (Laland and Brown 2002: p. 190).

Another questionable assumption of evolutionary psychology is that the human brain attained its current form in the Pleistocene, and has not significantly changed since then, especially not in the Holocene, or recent, period in which human culture has become so immensely complex. As Laland and Brown argue, this view rests on the presupposition that complex characters take a long time to evolve. However, there is no warrant for this assumption in the current state of evolutionary biological understanding. Indeed, there is some evidence that complexity can evolve quickly (Laland and Brown 2002: p. 190), a view which, as they point out and as we have already noted, Wilson maintained in the original edition of his *Sociobiology* (Laland and Brown 2002: p. 191) and continued to maintain in the course of the development of gene-culture co-evolution theory.

None of these critical points in themselves scupper the project of evolutionary psychology, for they all amount to the pointing out of complexities which have been ignored, rather than the propounding of an objection of principle to the whole enterprise. Laland and Brown argue that these complexities do not offer insuperable obstacles to the application of evolutionary thinking to the understanding of human beings. As they assert:

> There are rigorous methods for detecting the action of natural selection (Endler 1986), for isolating characters (Wagner 2001), for determining whether a character is an adaptation (Sinervo and Basolo 1996; Orzack and Sober 2001), and for drawing inferences about how characters have evolved (Harvey and Pagel, 1991)' (Laland and Brown 2002: p. 194).

However, their conclusion that evolutionary psychology has yet much to do in refining its understanding of evolutionary processes and devising more rigorous techniques for testing its hypotheses to meet the charge of 'Just-So-ism' seems to be well-merited.

In addition, the comparative neglect within evolutionary psychology of the role of cultural processes upon the selective environment, stemming from the idea that the human brain is both replete with domain-specific modules and has not significantly changed from the time of our Stone-Age ancestors, appears to be a significant weakness. The approach of gene-culture co-evolution, as championed by Edward O. Wilson and others, appears to be preferable in this regard, especially as Wilson has not committed himself to the view that human nature has remained unchanged since the Stone Age. He may be wrong about how much is currently changeable in human nature, but he is willing both to admit that evolutionary processes can be rapid, even with respect to complex traits, and that cultural processes have to thoroughly integrated into the story of human development, at both the individual and the species level.

Developmental Systems Theory and Dupré

One of the most trenchant recent critics of the project of evolutionary psychology is John Dupré. In this section we will examine his objections, as set out in Chapter 2 of his book *Human Nature and the Limits of Science*. He there proposes an alternative to the approach of evolutionary psychology, partly on the grounds that it is too gene centered an approach to evolution, and partly because it is said to operate with a concept of the mind which is open to serious objection. To what he sees as the gene-centrism of evolutionary psychology he offers the alternative which has become known as 'developmental systems theory'. To what he diagnoses as the faulty concept of mind he opposes a society-oriented view of mind deriving in part from the work of the later Wittgenstein. We will, therefore, examine these two arguments in this section.

Although Dupré's specific target in this discussion is evolutionary psychology, his criticisms have a more general relevance to the whole sociobiological project. He begins by making the following observations upon the whole project of sociobiology and its various sub-varieties, especially evolutionary psychology. Firstly, everything we do is something we have evolved the capacity to do. But the existence of the capacity may be thoroughly uninformative as to why, when or even whether we actually exercise that capacity. Thus, he argues, even if evolution can explain why people engage in some social behaviour at all, such as holding cocktail parties, it is still social facts (for example, the Bridge Club calendar) which provide the 'illuminating' explanation of why a particular party takes place (Dupré 2001: p. 23).

Thus, he argues, let us suppose that we are in possession of a maximally illuminating form of the explanation offered by evolutionary psychology – that is, one in which the following factors are all in play:

1 behaviour is directly caused by a specific structure of the brain;
2 the structure in the brain is caused by a specific gene/set thereof;
3 the genes are there because they were selected for their ability to produce the brain structure.

Even so, there would be much about the explanation of the behaviour which would require us to go beyond the modules – and look instead to social facts (Dupré 2001: pp. 24–5).

However, Dupré accepts that this first point does not in itself torpedo the whole project of evolutionary psychology. He accepts that even if the existence of numerous modules does not 'illuminate' much about actual behaviour, 'it would surely be interesting to know about them anyhow, and it would be plausible that such knowledge would contribute something to our understanding of actual behaviour' (Dupré 2001: p. 25). So he sets out to question 'the a priori arguments' underlying evolutionary psychology.

Before going on to consider Dupré's critical argument, we should note that his use of the term 'illuminating' in the observations just given begs important questions. What we find illuminating depends on what we want to know. We will usually know all the relevant social facts he mentions. Why not say that appeal to putative modules is illuminating because it explains why social facts of this kind are forthcoming, such as that people meet in social groups of a wide variety of kinds at which mood-altering drugs are consumed? This would be the case when we are in full possession of the relevant social facts of the kind to which he attaches such importance, but are still in the dark about just what is going on in that situation.

This might befall an anthropologist who is striving to grasp the role within a group's social life of some ritual that he has carefully observed and is struggling to come to grips with. Consider how the Trobriand islanders converted the game of cricket, which they had acquired from their British rulers, into an inter-group ritual of social cohesion. It ceased to become a competitive game (for the visiting team always 'won') and became a social ritual of hospitality and group solidarity (Weiner 1978). What this suggests is that the identification of what the 'social facts' are is relative to a background theory of what the behaviour observed is an instance of. Such background theory could well postulate a modular basis for the behaviour. If so, then a theory such as evolutionary psychology might well 'illuminate' what the social facts actually are.

Returning to Dupré's argument, he now undertakes to demolish the theoretical presuppositions of evolutionary psychology. He argues that though our genes did evolve over a very long time, and although the genetic basis for our brains has probably not changed a great deal since the Stone Age, the centrality of genes to evolution, and to the building of brains is typically greatly exaggerated, and the role of genes in biology does not require the view that we have 'Stone Age' brains (Dupré 2001: p. 26).

Dupré uses 'evolution' to mean 'any change over time in the distribution of properties across a population' (Dupré 2001: p. 26). That is, he doesn't restrict the term to 'change over time in the frequency of genes in a population'. Using this concept of evolution, he suggests, leaves it an open question how many of the former changes are to be attributed to the latter, although clearly he thinks that not many – or any – of the latter may be so explained.

Dupré now introduces a theoretical perspective that directly challenges gene-centrism. This is developmental systems theory (DST), first elucidated by Susan Oyama (Oyama 1985). He summarises the theory as follows (Dupré 2001: pp. 28–31).

From the point of view of the development of an organism, genes are only one of vast array of resources to be deployed. Other resources include: extra-nuclear chemicals/organelles in the maternal cell; the mother's reproductive physiology; environmental resources – obtained either from the environment directly or via parental construction; parentally deployed behaviour. These produce 'iterated cycles of development' and must produce organisms capable of assembling all the same kinds of resources for the next generation.

Evolution can only work on alterations that can be transmitted to future generations – genes fit this requirement, but are not the only things of this kind. As organisms become more cognitively sophisticated, genes become less unique in their transmissibility character. In particular, 'the human brain develops in ways that are enormously sensitive to the environment in which it operates'. Dupré claims that evolutionary psychologists are disposed to deny or minimise this point.

Hence there is no more sense in which 'genes build brains' than there is in the claims that 'wombs build brains' or 'schools build brains'. All these things are necessary conditions – not just the genes. If so, there is no reason to suppose that the brains built nowadays *are* the same as the brains built in the Stone Age

– genes aren't blueprints, specifying a unique building/structure at the end of construction (for more on this idea, see the appendix to this book).

DST is not, however, the 'blank slate' theory, excoriated by evolutionary psychologists, that brains are 'infinitely plastic' in response to environmental variation. Rather, DST makes plausible idea that 'human brains evolve' (in Dupré's sense of the word –see above) 'at the speed of cultural change'. However, in spite of this superior standpoint, he admits that DST has yet to produce a program with achievements 'comparable to orthodox neo-Darwinism'.

What response should a defender of evolutionary psychology make to this argument?

Firstly, let us consider his claim that because DST includes non-genetic, and non-biological, elements within the developmental matrix of the human brain, this theory implies that the brain may be said to 'evolve at the speed of cultural change'. Clearly, the governing thought here is that if the human brain does evolve with culture, and cultures are so varied in time and place, we must abandon any hope of discovering a universal form to the human brain, of the kind which evolutionary psychology committed to establishing. The specific form of this hypothesis for evolutionary psychology is, as we have seen, that the brain has a modular structure, common to all human beings at all times and places. However, in reply to this argument a defender of evolutionary psychology would have some grounds for claiming that DST seems to leave that issue open.

That is because, firstly, the possibility is left open by DST that human brains do always need to develop some modular structure or other even if, given that genes are only necessary, and not sufficient conditions for the development of brains, it is possible that different non-genetic factors, especially cultural ones, will produce brains with different modular structures. Something like this hypothesis is in fact at the heart of gene-culture co-evolution theory, as we will discover in the next chapter. Thus, the question of whether any normal human brain has a modular structure is logically independent of the question of whether they all have the same structure. Secondly, it may still be that what is selected is a uniform human nature, operative in a variety of situations, because the elements of that nature are established by certain universal demands placed upon the human organism by all environments, in spite of the different cultural matrices in which individuals come to maturity.

There seem to be various ways of envisaging this possibility. One is that cultural variability at one level may mask a uniformity of effect at a more fundamental level. For example, perhaps normal human development requires human children to receive the support and love of some male adult or other. However, different cultures may assign this role to different categories of male adult – wife's brother, rather than biological father, to take an oft-cited example. How different cultures actually are from each other is not necessarily a straightforward matter to determine. It is a matter of classification, and thus we need to ensure that we are operating at the right level of classification.

This is a problem with which some anthropologists have been wrestling as they attempt to bring the search for universals into contact with the prevailing ethos of their discipline, which for much of its existence has had a preoccupation with the unique and the exotic (from the point of view of western culture) (see, for example, Brown 1991: pp. 1–6). The problem that all biological theories, whether evolutionary psychology or DST, have when they come to consider the human species is that the classification of the elements of human social life is, at least initially, infected with the indeterminacy of the classifications prevalent in the social sciences and humanities. In those disciplines what may be a vice has, perhaps for reasons of self-preservation of those disciplines, been turned into a virtue.

It is part of the hope of the sociobiological project, including evolutionary psychology, that the dialectical interaction of biological theory and sociological observation may in the course of time lead to a theoretically robust form of conceptualisation and classification. As we will be observing in chapter five, this is what consilience seeks to achieve. Hence, the introduction of cultural elements into the idea of the 'life-cycle of human beings' which is to be selected by evolution does not in itself settle the question of whether or not the result of that selection upon those life-cycles is to produce a uniform human nature. How we classify those cultural elements is going to be a crucial matter, and evolutionary psychology may find possibilities for the production of the necessary uniformities, modular or otherwise, within a well grounded set of classifications.

In any case, when one examines the kinds of human capacity that have been thought by supporters of evolutionary psychology (and not just them) to be universally present in all normally-developing members of the human species, one is struck by how basic to all forms of human life they often seem to be. Such a basic set of capacities may nevertheless show variation between human beings and groups, but it seems to be unnecessary to suppose that there must be such variation just because there is variation at some fairly superficial level.

Language is the clearest case. To cast the matter in the modular hypothesis favoured by evolutionary psychology, one can suggest that if there are modules essential to the possession of the capacity for a natural language, then the genes which build these (agreed, together with other causally-necessary elements) will presumably do so in a standard way. Otherwise we would have to suppose that the brain modules which enabled stone age people to speak are different from those which govern speech in contemporary people. Is there any reason to suppose this? Granted, the specific languages are now different, and the things people talk about, the concepts they possess will be very different (though evolutionary psychologists claim all human beings need to possess a module-provided set of basic concepts, such as 'inanimate object', 'living being' and so forth – see Pinker 1997) Thus, if there are modules, then it looks as if their presence will be necessary to a wide range of human behaviour – and may constrain it.

In order to show that human brains at different times and places are likely to be different structures, Dupré needs to show more than that different brains in these circumstances come to possess different concepts, norms, and so forth. He needs very precisely to show that they must come to possess different modules (or differ fundamentally in some other respect), and/or that modules are not necessary (or necessary for very much).

Of course, the brains may still be different if the genes for certain modules do not become operative in different circumstances. Perhaps modern people's brains consist of very different modules from those of their ancestors, although they possess the same genes, because different module-constructing genes come into play in different environments, or because the same genes produce different outcomes in differing circumstances. But it seems an open question whether or not this is so, and one that cannot be settled simply by pointing to the existence of different cultural elements in the brains of members of different societies.

Evolutionary psychology seems most persuasive at first glance in those areas of human psychology which are the most basic to the functioning of the human organism – language, basic recognition of objects in space, human beings, other living creatures, etc. It is here also that the modular picture of the mind seems easiest to establish. This is because, firstly, it is very difficult to see how a mind initially devoid of these basic component concepts could acquire them on the basis of general intelligence plus experience.

Secondly there seem to be many empirical cases currently available for the investigation of human beings with certain kinds of localised brain damage whose basic mental capacities in these areas are also impaired in highly specific ways. This seems to permit the construction of the specific modular hypotheses without reference to any putative problems in the EEA to which the modules would provide their possessors with a solution and which would thus account for those modules' selection by the environment.

In other areas of human psychology, however, it has to be admitted that evolutionary psychology has its work cut out to establish its case. Examples of such modules are morality and sexual behaviour. Here it can be argued with apparent plausibility that there are simply no uniformities underlying surface differences comparable to the language example. At any rate, evolutionary psychologists have to work harder to establish that there are such uniformities. Also, it seems much harder to show that sexuality and morality are analysable into discrete components that are interrelated in some rule-governed manner as one can show in the language case. This makes the very detection of uniformities much harder, since there may be no very clear way of delimiting the domain in question. One may also press the question of whether there are examples of localised brain damage correlating with specific disabilities – or at least deeply altered patterns of behaviour and tendencies to act – in these areas.

Reverting for a moment to the DST, It is pertinent to note that this theory, while providing telling reasons for making genes into necessary, but not sufficient, factors in the development of the traits in phenotypes, may not have succeeded in showing that there is no important difference between genes and other factors in phenotype development. As Sterelny and Griffiths note (1999: p. 109, citing Agar 1996; Sterelny, Smith and Dickison 1996), one powerful counter-argument is that replicators, of which genes are a clear example, are selected because of their role in ensuring that offspring show certain kinds of similarity to their parents. Replicators may thus be said to contain quasi-semantic information about the phenotype – their function is to ensure that the phenotype does end up by conforming to a certain pattern.

This function is not shared by the non-replicator elements in the developmental matrix of the phenotype, even though their presence too is a necessary element in phenotype development. Sterelny, who champions this view, does, however, point out that this argument applies to all the replicators operative in phenotype development, including the copying mechanisms which are not genes. In sum, then, DST may adopt too holistic an approach to phenotype development, one that obscures important differences between different elements in the developmental matrix.

Returning to Dupré's critique of evolutionary psychology, he next focuses upon the concept of the human mind which he believes that evolutionary psychology is committed to, and seeks to show that it is untenable. Its supposed tenability is what gives the project of unearthing a uniform human nature its apparent plausibility. Dupré begins by asserting that the 'belief/desire' model of 'folk psychology' is 'a roughly correct model for the explanation of behaviour'. He considers the claim that a properly scientific account of behaviour can only be got from this model if the beliefs and desires 'can be identified with properly physical states of the brains that are the real physical causes of the bodily movements that constitute the action to be explained' (Dupré 2001: p. 32).

In reply to this claim, Dupré argues against the view that the mind is to be identified with the brain, and indeed argues against the view that the mind is a kind of 'thing' at all. To establish this position he offers the following argument (Dupré 2001: p. 33):

1   it is impossible to characterise a human mind without appeal to language: even if thought does not always require language, some kinds of human thought are impossible without language;
2   a language is not a property of an individual but of a linguistic community;
3   thus it is not possible to characterise a mind without some (implicit or explicit) reference to a community of which the possessor of that mind is a member;
4   thus, the kind of thing a mind is (its ontology) depends on how aspects of mind are embedded in a community;
5   since brains are not thus ontologically related to communities, they cannot be identified with minds (Dupré explains that although brains are *causally* dependent on their social context, it is *logically* possible for brains to exist independently of that context, which is not true of minds (p. 34, fn 11)).

In support of (1) Dupré argues that humans are self-conscious; they are aware that they represent the world to themselves. This involves awareness of a distinction between 'the way X represents the world' and 'the way the world is'.

This implies awareness of representations, which for human beings are linguistic entities. So human self-consciousness requires awareness of such linguistic modes of representation (Dupré 2001: p. 34)

In reply to this argument, one may suggest that it does not really get us very far. For one thing, there is good reason to suppose that some other mammals – pigs, chimpanzees and dolphins have all been put forward as candidates – possess self-consciousness. Yet the question of the degree to which they can be said to posses language is still unresolved. There do also seem to have been cases (such as Victor, the wild boy of Aveyron) of human beings who possess self-consciousness but not language (for a discussion of this case, see Ridley 2003: pp. 169–70). It may be that these examples are not about the mode of self-consciousness to which Dupré refers in (1) above. Here, self-consciousness is defined in terms of awareness of representations. But if this is how Dupré's point is taken, then it becomes tautological – simply, and uninformatively, asserting that a being which is aware that it is using linguistic representations is necessarily a language-user. We may go on to argue that we should explicate the concept of 'possessor of a self-conscious mind' in such a way that it is left an open question how far such an entity is conceptually tied to the possession of language. Otherwise we may miss key levels of explanation out of our theory of mind.

In support of points (2)–(5) Dupré offers the key element in Wittgenstein's private language argument (Wittgenstein 1953: sections 243–315). 'Meaningfulness … what makes a usage right or wrong, depends on the existence of norms, or rules, which is to say practices with normative force in a community' (Dupré 2001: p. 35). He reaffirms the strong claim that 'the possibility for self-consciousness is a possibility only for a being embedded in a linguistic community' and rejects the picture theory of meaning. There then follows a version of Winch's famous Wittgenstein-derived argument (Winch 1958). Movements of the hands only count as the writing of a cheque in the right social context (Dupré 2001: p. 36). In general, human beings construct for each other possibilities for behaving which would be impossible without such construction – for example, lifts for the wheelchair-bound to ascend high buildings (Dupré 2001:pp. 36–7). Also, people come to maturity in various kinds of society and may not be able to adapt from one to another – they learn very specific ways of behaving. (Dupré 2001: p. 37). The conclusion of these points is the claim that human beings depend ontologically on human social contexts. They become the specific human beings they are as a result of their embedding in specific contexts. Dupré denies that this is the 'blank slate' argument, and claims it is a point which is 'constantly … lost on those who wish to provide universalising accounts of human nature' (Dupré 2001: p. 37).

How might a defender of evolutionary psychology reply to this argument? On the face of it none of these points should be a problem for evolutionary psychology, for they all apply to language, which is arguably the strongest case for the modular nature of the mind. There is no obvious reason why defenders of evolutionary psychology cannot accept the following claims, and indeed rely upon them as a key part of their argument. Culture supplies the specific sounds and concepts that make a specific language what it is. If culture didn't do that, certain ways of thinking would not be possible. People find it difficult to learn a language other than their first one. Uttering sounds only counts as speaking a language in certain social contexts. There is an open-ended set of linguistic possibilities, as exemplified by the 6,000 or so existing human languages. These all seem compatible with the modular nature of the linguistic capacity of the human brain. This modular nature appears to be universal across human beings.

The difference between Dupré and evolutionary psychologists appears to come down to the issue of how constraining modules could be. The modular nature of language tells us nothing about the limits of what can be expressed in any language. The existence of a finite set of syntactical and semantic rules is both compatible with, and makes possible, an open-ended set of specific sentences within a language; and the existence of a Chomskyian 'deep structure' is compatible with and makes possible, an open-ended set of specific languages. Dupré focuses on the open-endedness; evolutionary psychologists focus on the finite structures. Are they, then, talking past each other?

Well, Dupré sometimes seems to suggest that he thinks there are no finite structures (though its not clear that he does think this; sometimes his point is rather that even if they exist they cannot explain very much of importance about human behaviour). He also seems worried about the idea that 'structures impose limits on behaviour'. The unspoken worry may be the one we addressed at the end of the last chapter, that such constraints might make certain kinds of desired human society impossible for human beings to attain, and the connection with evolution seems to raise the prospects of Social Darwinism. But the analogy with language suggests that structure/limitations in one dimension are compatible with, and make possible, open-endedness in a different dimension.

What this suggests is that we need a careful statement of what structures and constraints are necessary parts of any specific modules, and what forms of open-endedness (if any) they make possible. The existence of modular structures for various forms of human interaction in, say, human moral and sexual behaviour may make possible an open-ended set of variations on the theme of human institutions governing, say, marriage and property ownership. But structures will certainly make certain conceivable forms of human interaction not easily or even at all attainable. It is not possible a priori to say what these might be. Dupré has not yet given reason for rejecting the evolutionary psychology project, unless he can demonstrate that there are no structures underlying human behaviour. The simple fact of pointing to endless variety is in itself no argument, for, depending on how those variations have come about, their existence is compatible with, and may require, underlying structures. And his arguments concerning the indispensable role of social context is, as we have seen, perfectly compatible with the existence of the structures.

The final points that Dupré makes concern other difficulties involved in evolutionary theorising and in the conceptualisation of the modular hypothesis. The first point concerns the difficulty of engaging in speculation about what features of organisms are adaptations – that is, features selected by the environment of the organism's ancestors and passed on to the organism as a genetic endowment. The problem here, Dupré argues, is that evolution only needs good enough, not optimal, solutions to environmental challenges. So we are unable to speculate what solutions evolution would be likely to throw up, since any of a countless number of sub-optimal ones might have served to confer a reproductive advantage for an organism in a specific environmental context. What gets passed on to a descendant may, thus, be a sub-optimal adaptation, or something with no adaptive import at all.

This point is telling, but if it is regarded as decisive than evolutionary explanation appears to be impossible – even the modest amounts of evolutionary explanation which Dupré seems happy with. What it rather suggests is that any deployment of the idea of adaptation cannot be a simple matter of telling a plausible adaptationist tale (a 'Just So' story). Such tales can at best suggest hypotheses which might only become plausible (and perhaps never more than that) when they are interrelated with other hypotheses and data in a complex way. They might still be the best explanations even if they are irredeemably speculative. After all, fallibilism in all areas of science means that no theory and associated pattern of explanation ever loses a speculative cast entirely.

The second difficulty that Dupré raises is to ask how the thousands of modules supposed by evolutionary psychology to constitute the human brain interrelate. How is it determined which gets priority? Here he raises the metaphysical spectre of the controlling homunculus required to determine interrelations between modules. This objection looks to be amenable to some thesis of complex interactions, as with any complex structure. After all, the other component organs of the human body interact and explanations of that interaction do not require the postulation of a homunculus. With enough complexity in the rules governing interactions between modules – assigning priority to certain modules in certain complex situations, and to others in other such situations, there need be no necessity for a single supermodule to do any final determination. This looks to be an area where something like Hayek's concept of 'spontaneous order' will be useful (see Hayek 1960: pp. 148–61).

The conclusion that we thus reach after this consideration of Dupré's critique of evolutionary psychology is that, at worst, it poses difficulties, rather than offering a decisive refutation of the whole project. The distinctive idea of evolutionary psychology, the modular theory of the mind, may turn out to be false, or very limited in scope. However, the one area where it looks most promising, namely in the area of language, is the one in which Dupré makes what he thinks to be his most telling objections. Yet, as we have seen, those objections appear to raise no difficult issue of principle for the modular hypothesis.

They show that language has a necessarily social dimension. In that respect they show that one possible way in which language might have been set up in the human brain – so that all human beings develop the same language (syntax and semantics) as a matter of automatic development – is not what actually happens.

As we will see in the next chapter, a version of this possibility is considered and rejected as an evolutionary possibility by Lumsden and Wilson in their development of gene-culture co-evolution. Such a hypothetical single-language option is difficult to fathom precisely. For, of course, even if the syntax of a natural language remains constant, the vocabulary certainly does not. If it is to be useful in enabling human beings to adapt to changed circumstances, it cannot remain constant. And if vocabulary changes, then there must be the possibility of different changes being made by different sub-groups of speakers. This will then give us in due course more than one language. This, after all, is a bit like what occurs in language families, such as the romance languages.

So there are good evolutionary reasons why language should permit social input into language formation in the individual. The single-option language might have developed in a universe in which no significant environmental changes ever took place. But such a universe is not ours, and it is difficult to see how it could be a universe with any degree of complexity whatsoever.

Chapter 4

Gene-culture Co-evolution

Let us now consider gene-culture co-evolution, a version of which, as we have already noted, Wilson began to develop in the late 1970s along with Charles Lumsden, partly in response to the claim that human sociobiology as hitherto conceived was insufficiently sensitive to the cultural dimension of human development. In their preface to the book, the authors note that sociobiology had not yet paid sufficient attention to the human mind itself or to the fact that human cultures are so very diverse. They deny that sociobiology as an approach suffers insurmountable difficulties of an epistemological kind, or that it is inherently a politically (and by implication morally) objectionable mode of thought. However, they do accept that, as applied to our species, sociobiology must take full account of human beings as cultural, in a very distinctive and full sense of that word and claim that it is the failure of sociobiology to do this that has underlain much of the hostility to the theory (Lumsden and Wilson 1981: p. ix). In this they may have been taking too optimistic a view, given that hostility to sociobiology of any kind remains strong in certain culturalist circles, as we will be seeing later, in spite of their concerted effort to work the specificities and uniqueness of human culture into their theory.

It is important to note that they were matched in this effort by two other theorists, Cavalli-Sforza and Feldman, who published *Cultural Transmission and Evolution* in 1981, the same year that Lumsden and Wilson published their work. The theory of gene-culture co-evolution thus had more than one set of exponents right from the start, and the view championed by Lumsden and Wilson has not had the influence which the impressively comprehensive, if avowedly speculative, theory they produced might have led them to expect.

Gene-culture co-evolutionists ('co-evolutionists' for short), combine a belief that culture evolves as it is passed on between individuals with the view that an individual's genetic constitution may influence, positively or negatively, which elements in the culture pool that individual may adopt. However, culture is also viewed as serving to modify the selection pressures which work upon the gene pool. Hence, influence cuts both ways. (Laland and Brown 2002: p. 243).

To move on from these general positions to a more detailed examination of these complicated interactions, it is necessary to analyse culture along the same lines as genes. Lumsden and Wilson used the term 'culturgens' to pick out discrete elements of human culture which could be passed on between individuals, mutate, and exercise a selective effect on human genes. At the level of the individual, the absorption of a culturgen was held to be a function of 'epigenetic rules' produced by biology and the learning processes activated within the individual by social interaction (Laland and Brown 2002: p. 244). The aim of this theory was to produce models with some predictive power concerning how the culturgens and the genes would alter across time and cultures. Lumsden and Wilson concluded that genes would probably evolve that predispose the individual to absorb some culturgens rather than others, that such biases might be intensified by social conformism and that the speed of genetic mutation could be altered in either direction by the influence of culturgens (Laland and Brown 2002: p. 244).

Having thus sketched in the general position adopted by Lumsden and Wilson, let us now examine it in greater detail before considering rival versions of the idea which ascribe rather more autonomy to the cultural realm than theirs does.

Lumsden and Wilson's Version of Gene-culture Co-evolution

It is striking that right at the start of their discussion Lumsden and Wilson are careful to differentiate the theory they are going to develop from the crude but popular notion that it makes sense to speak of genes that direct or prescribe behaviour. They are willing to accept that sociobiologists and their ethologist forebears have tended to speak in these terms. However, they suggest that, in the human case, at least this is not a defensible view: 'Behaviour is not explicit in the genes' (Lumsden and Wilson 1981: p. 2).

The alternative view bears some resemblance to aspects of evolutionary psychology, in that it focuses on the idea of mind/brain structures that underlie and partly explain behaviour without fully determining it. The role of genes (which have to be understood in the full range of their complexity – see the Appendix of this book for further details) is to produce a set of biological processes which Lumsden and Wilson dub 'epigenetic rules'. These rules (ERs for short) direct the 'assembly of the mind', but do so in a subtle and complex way by responding to inputs of information from the physical environment and the cultural context of the mind in question. The mind of a human being, therefore, is structured in the course of its development by the interaction of gene-based processes and the external, including cultural, context within which that development takes place. Of course, this culture is itself ultimately a product of minds and is understood by Lumsden and Wilson as the conversion of ERs in individual minds into 'mass patterns of mental activity and behaviour' (Lumsden and Wilson 1981: p. 2). The authors, therefore, aim to give full recognition to the ability of the human mind to be free ranging, rather than tightly constrained by genetic processes, and to be creative of a wide variety of cultures.

Lumsden and Wilson aim to locate the characteristics of human culture within a broader account of the evolution of mentality amongst the organisms which can meaningfully be said to possess this feature. Some species, such as the other great apes, have developed culture-like forms of behaviour that are dubbed 'protocultures'. They are capable of the activities of simple learning, imitation and teaching and can create varieties of tradition within discrete populations. The key step forward from these capacities which was taken by human beings is the mental activity of 'reification', which is Lumsden and Wilson's term for the construction of symbols and other abstract representations of the world. With reification we have the emergence of what they term 'euculture' which, as far as can be judged from the historic record, is unique to human beings and perhaps their near, but now extinct, relatives (Lumsden and Wilson 1981: p. 3). However, the authors reject the restriction of the concept of culture to the creation and manipulation of the purely symbolic, partly because it would involve expelling from the concept of culture many forms of more primitive mental activity, such as imitation, which are never wholly superseded within euculture and from which symbolisation probably emerged, and partly because it produces a question-begging isolation of the human case from that of other species (Lumsden and Wilson 1981: p. 4).

Reification is what allows human beings to produce concepts with which they can classify and reclassify features of their world, think about them, formulate hypotheses with respect to them, and even produce what Lumsden and Wilson call 'mentifacts' such as gods, spirits and totems. These are concepts having no real basis in experience at all, but which may serve other, deep, socio-psychological purposes, such as the binding together of human groups (Lumsden and Wilson 1981: p. 5). They also note a phenomenon which has struck other theorists of human cognition, and been the subject of extensive critique (see, for example, Plumwood 1993: pp. 41–68) which is that human beings often resort to a binary logic. That is, they divide what is a continuously varying reality into two discrete, and often opposed, or at least contrasted, groupings, such as ingroup/outgroup and child/adult, and police the frontiers so created by rites of passage, forbidden interactions, and so forth. (Lumsden and Wilson 1981: p. 5).

Much of what the authors say here with respect to the role of reification is uncontentious, forming part of the received wisdom about the importance of language in the development of the human mind and the creation of culture. But it is perhaps useful to note that they fully accept this orthodoxy and attempt to account for it within the terms of their gene-culture co-evolution theory, rather than to explain it away. They are thus not seeking a form of eliminative genetic reductionism.

Let us turn now to consider in a little more detail how the gene-culture interactions are supposed to take place. The key idea here is that of a programme of individual development. As we noted above, Lumsden and Wilson accept that earlier versions of sociobiology moved from genes directly to behaviour in a way that completely ignored the whole business of individual development. But to understand how genes and behaviour interconnect a developmental account is crucial.

To construct an adequate account, however, we will need some basis on which to analyse culture into discrete units so as to investigate the interaction between culture and genes in the construction of epigenetic rules. As we have noted above,

Lumsden and Wilson employ the concept of the 'culturgen' for this purpose. They explain that a culturgen is a 'relatively homogeneous set of artefacts, behaviours, or mentifacts' (Lumsden and Wilson 1981: p. 27) which are formed on the basis of their having some common feature, or having an overlapping set of such features, in accordance with Wittgenstein's concept of family resemblance (Wittgenstein 1953). The authors argue that such a conception is widely accepted amongst archaeologists who have to select or formulate discrete artefacts and assemblages for study. Archaeologists use the concepts so formed to trace the history and evolution of artefacts, and thus the history and evolution of cultures, in objective ways. That is, the ways in question are such that the criteria used to form the classifications are not subject to further personal evaluation once they have been formed.

This comparison with archaeology ought at least to show that, if there is a problem with digitalising culture in order to make the gene-culture co-evolution theory work, then this is a problem which is not restricted to that theory, but arises in any attempt to study human culture systematically, including the social sciences. We will return later to the issue of whether or not culture can be digitalised when we have considered other versions of gene-culture co-evolution theory. However, It is a useful corrective at this stage to a possibly oversimplified picture of culturgens to note that the authors themselves recognise that some culturgens are easier to individuate than others. They argue, however, that this fact alone does not automatically count against the usefulness of the concept, for there are other examples of concepts in biology, such as that of a species, that are easier to apply in some cases than in others, but that nevertheless are indispensable to an understanding of biological phenomena (Lumsden and Wilson 1981: p. 30). This, they say, suggests that the best research strategy with respect to culturgens is to begin with cases where objective individuation is most readily achieved and use them as the starting point from which more complicated and diffuse cases may be examined.

Having introduced the concept of the culturgen, we are now in a position to follow Lumsden and Wilson's account of gene-culture co-evolution. It will be recollected that, on their account, in the course of human development genes interact with the environment of the human organism, physical and cultural, to produce the ERs. These are not simple entities, but comprise the components of complicated sequences of events that take place at different places within the nervous system.

Lumsden and Wilson divide ERs into two broad classes. The first, the primary ERs, are subject to greater genetic control than those in the other group, the secondary ERs. The primary ERs are chiefly employed in the sense modalities, encompassing such phenomena as the way human beings classify colours. The point about ERs is that they constrain the human mind to work in some ways rather than others. Human beings across the whole range of cultures that have been investigated turn out to classify colours in very similar kinds of ways. There are differences between them, but the differences are systematic, not random as would be expected if such classifications were determined entirely by cultural factors (Lumsden and Wilson 1981: pp. 43–8).

The secondary ERs are important for the more advanced forms of mental activity, such as decision-making and the evaluation of features of the world around us. Their role is that of disposing individuals to transmit certain culturgens rather than others. The ERs which operate in these forums encompass a variety of phenomena that have already been unearthed by psychological investigation, such as 'fuzzy logic', whereby predicates are treated as approximate, rather than simply true or false, and satisficing, whereby human beings act so as to reach some acceptable level of attainment, rather than adopting a maximising or optimising strategy (Lumsden and Wilson 1981: p. 96).

From the variety of psychological studies which have examined secondary ERs, Lumsden and Wilson extract various generalisations, such as the following. Firstly, it is during early childhood that secondary ERs have their greatest rigidity. But however rigid they are, there is nothing in the human case which matches the form of imprinting which is easily found among other animals. This is because genetically controlled learning in human infants takes too long to qualify for that label. Again, the more that environmental circumstances have an important part to play in determining whether or not a category of behaviour has an effect on genetic fitness, the more likely it is that that category will be assessed by the conscious mind, and the more flexible the behaviour is likely to be (Lumsden and Wilson 1981: p. 96).

However, in case it should be thought that such flexibility is automatically evidence of pure cultural transmission of culturgens, with no biological input at all, the authors hasten to add that flexibility of this kind is still compatible with the existence of core rules of cognition and decision-making that are themselves genetically determined. That is, flexibility can just as well be understood as a genetically prescribed strategy for those cases in which sensitivity to the specific environment in which it occurs is crucial for the adaptiveness of the behaviour. However, they accept that whether or not there are such rules is a matter for empirical study to unearth. They might not exist (Lumsden and Wilson 1981: p. 97).

When culturgens enter the mind, they are processed through a sequence of primary and secondary ERs. The authors speculate that in practice many, or perhaps a majority, of cases of cognition and behaviour will involve the interactions of various primary and secondary ERs (Lumsden and Wilson 1981: p. 52). This processing is a matter of determining the probability that one culturgen will be used rather than another by the mind in question. Thus, ERs can be thought of as establishing a certain form of bias in the mind – some culturgens are more likely to be used than others, and this differential probability can be expressed mathematically in a set of usage bias curves (Lumsden and Wilson 1981: p. 7).

Theoretically, the authors argue, culturgens might be subject entirely to genetic constraint, so that everyone transmitted (used and passed on) exactly the same culturgen on every relevant occasion. This is a version of what is usually termed instinct or a fixed action pattern. Or they might be subject to pure cultural transmission, where no inherent bias for the transmission of culturgens can be detected. There would be no greater probability that one rather another culturgen would be transmitted. Both of these extreme possibilities are said to have their serious downside as accounts of culturgen transmission in human beings. The former, if it is to allow for the euculture with which we are familiar, would require brains of a size and complexity which it would be beyond the scope of DNA to construct (Lumsden and Wilson 1981: p. 342). The latter would require a different kind of unattainable physical substrate, namely a continuously evolving genome which encodes for a whole battery of procedures for fine-tuning decision-making so as to bring out the relatively stable patterns we discern within a range of choices which are at each point completely open and continuously changing (Lumsden and Wilson 1981: pp. 10–11). Additionally, our sensory apparatus is clearly subject to limits of acuity and range of sensitivity, which means that it is not able to respond equally to all the possible ranges of sensory stimuli. It must be biased in certain directions to function effectively at all.

Further, there are arguments from evolution itself. Any non-eucultural ancestors of an eucultural species must have evolved biased forms of mentality, such as prepared learning, in order to survive. Those forms of bias will persist into the eucultural phase. Also, any species which had somehow adopted a completely open, unbiased set of ERs would be bereft of means for distinguishing between adaptive and non-adaptive culturgens. Such a species would be subject to invasion by mutants that had evolved ERs biased towards the adaptive culturgens. The unbiased ER choice would thus not be an evolutionarily stable strategy.

This leaves the third possibility as the one easiest to reconcile with the known facts of human development, namely gene-culture transmission. The definition of this is 'transmission in which more than one culturgen is accessible and at least two culturgens differ in the likelihood of adoption because of the innate epigenetic rules' (Lumsden and Wilson 1981: p. 11). In other words, the ERs bias the mind to accept and pass on certain culturgens rather than others of those that are available for such transmission. This, the authors explain, is the reality behind the idea that gene-based natural selection works so as to 'keep culture on a leash'. But they are at pains to emphasise that this idea does not amount to the rigidly determinist notion that genes create hard-wired algorithmic processes. Rather the 'leash' is metaphorical expression for certain biases or tendencies in the human mind (Lumsden and Wilson 1981: p. 13).

They also suggest that for a variety of reasons a species with euculture will experience a lengthening of the leash – a weakening of genetic constraints. Firstly, the maintenance of a highly sensitive neural system necessary to impose tight constraints on culturgen choice comes at too high a metabolic cost. An eucultural species develops too many choices to be tracked and filtered entirely by an algorithmic process. Secondly, the culturgen set within the mind of a member of an eucultural species clearly contains an innovative, dynamic element. A key aspect of the adaptive value of culturgens for such a species is the possibility for new developments which lie within them and which will lie dormant if the choice of culturgens is tightly specified by genetic factors (Lumsden and Wilson 1981: p. 13). It is worth once again highlighting the way that the authors are sensitive to the flexibility and autonomy of culture in eucultural species, while at the same time relating cultural processes to underlying biological factors. Their theory may in the end be still too weighted towards such factors, but they cannot fairly be accused of seeking a greedily reductionist view of culture.

What now needs to be grasped is the possibility of gene-culture co-evolution. The ERs can change as the result of genetic changes; the culturgen frequencies can change as the result of ER changes; or both can happen together (Lumsden and Wilson 1981: p. 11). The authors need, of course, to explain the mechanism whereby differential choices of culturgen transmission made by different individuals at a certain point in time can produce changes in the genotype prevalent in later generations so as to code for ERs which bias culturgen choices in a new way. To this end they claim that they need to assemble four types of evidence in support of the claim that eucultural species such as human beings engage in such coeveolution.

The evidence, of course, is what one expects any account of natural selection to contain. Firstly, there must obviously be evidence that there are indeed ERs of the kind they have described, which bias culturgen choice and which are susceptible to the sort of analysis needed to test the hypotheses of co-evolution theory. Then there must be evidence that there is in fact variation in ERs between different members of human populations. The authors claim that there is indeed abundant evidence of this. Such procedures as pedigree analysis and twin studies reveal 'evidence of genetic variance in virtually every category of cognition and behaviour ... including some that either constitute epigenetic rules or share components with them' (Lumsden and Wilson 1981: p. 16).

Then, if evolutionary theory is to get any purchase on the phenomena of euculture, it must be shown that cultural practices can have an important effect upon genetic fitness, that is upon the ability of individuals to pass on their genes to new generations. The authors claim that there is indeed such evidence, citing the evidence of certain agricultural practices which have enhanced or diminished the genetic fitness of the populations which engage in them (Lumsden and Wilson 1981: pp. 17–18). Finally, to get from genes to ERs, we need evidence that there are mechanisms that operate at the level of molecules and cells that convert genetic information into cognitive development. Here too the authors claim that there exists enough current evidence that these mechanisms exist, citing, for example, the activity of neurotransmitters that can have a marked effect on mood, concentration and social behaviour (Lumsden and Wilson 1981: p. 18).

Having demonstrated to their satisfaction that the kinds of evidence needed to sustain the co-evolution theory do in fact exist, the authors go on to make some interesting and important claims about the possible course of such evolution. The first is that even very small changes in the rate at which natural selection acts on individuals via ER constraint on culturgen transmission can lead to genetic changes across whole populations, biasing its members towards the favoured culturgen.

In turn, only a small change in such biases can lead to very striking changes in overall patterns of social organisation, as evidenced by Schelling's study in *Micromotives and Macrobehaviour* (1978). This process was noted by Wilson in his book on sociobiology, where it was referred to as the 'multiplier effect' (Wilson 1980: p. 9). The authors refer to the key process here as 'gene-culture translation', defined as 'the biological feedforward through the epigenetic rules of individual development to the formation of social patterns.' (Lumsden and Wilson 1981: p. 179). Here the theory connects up with the accounts familiar from social science of social processes of exchange of information, imitation, and distribution of information that account for the transmission of culturgens across social groups. The social level can itself have a direct effect upon the probability that a given individual may accept a given culturgen. If a certain proportion of members of a society has already accepted a culturgen (the precise proportion may vary from case to case) then that in itself will increase the probability that a given individual will accept it. This often achieves a simplification of decision-making for the individual that may have adaptive value (Lumsden and Wilson 1981: p. 180). Such tendency to conform to group norms can, of course, nevertheless be a strategy that derives from an underlying core rule of cognition.

Finally, and most strikingly of all, the authors suggest that changes of this kind can take place very quickly. Such genetic shifts, they argue, can become detectable in as few as 10 generations, or 200 years, and they later claim that 'the alleles of epigenetic rules favoring more successful culturgens can largely replace competing rules within as few as fifty generations, or on the order of one thousand years in human history' (Lumsden and Wilson 1981: p. 304). This, of course, implies that significant forms of genetic evolution within the human species with respect to the brain have taken place since the beginning of recorded history and that they are still taking place. This seems to set off Lumsden and Wilson's co-evolution theory from that of the evolutionary psychologists, who hold that the basic structures of the human brain were established during our long Pleistocene hunter-gatherer phase.

This theory, then, is compatible with rapid historical change of the kind which humanity has experienced in the last few hundred years, and which is such a contrast with the much slower rates of change over the previous phase of human history. The flexibility of many ERs, the innovation inherent in human culturgens, the large effects of small changes, and the potential speed of evolutionary change all permit rapid and wide-ranging changes to occur. But, while there have been many such changes, there is also continuity too. As the authors suggest at the end of the book:

> We should keep in mind that most of the wondrous inventions of science and technology serve in practice as enabling mechanisms to achieve territorial defense,

communication of tribal ritual, sexual bonding, and other ancient sociobiologic functions. (Lumsden and Wilson 1981: p. 360)

Returning to the account of co-evolution, we need to note some further key elements in the conceptualisation of the co-evolutionary process. Firstly, 'genetic assimilation' is the term used to describe the process whereby a change in the environment of human beings permits the survival of a wider range of individuals (phenotypes) than previously. If these phenotypes are at a genetically selective advantage, then there will be an increase in the number of genotypes that tend to produce such phenotypes. Depending on the precise course of evolution, the whole population may come to possess the predisposing genotype. This process is known as the Baldwin effect, and it is important to see that it is not an example of Lamarck's hypothesis of the inheritance of acquired characteristics. As Dawkins has explained, in the Baldwin effect '[l]earning does not imprint itself onto the genes. Instead, natural selection favours genetic propensities to learn certain things' (Dawkins 2004: p. 401fn). It needs to be emphasised that this process of genetic assimilation is avowedly a speculative one, at least for the behavioural traits in which co-evolution is most interested. However, the authors claim that it is a process that has been documented for human anatomical and physiological traits (Lumsden and Wilson 1981: p. 21).

Secondly, 'culturgen assimilation' is the process whereby cultural innovations tend to develop so as to realise all the possibilities permitted by existing ERs. The hypothesis that the authors offer is of rich diversities of cultural attainment clustering around those nodes that are favoured by the ERs. Those who are critics of biological approaches to human beings focus on the diversities, but fail to detect the underlying clustering directed by the ERs. The description that the authors offer of co-evolution, using these two concepts, is that culturgen assimilation follows genetic assimilation. That is, a few selectively advantaged culturgens (advantaged, it will be remembered, by change in the environment, which may itself be altered by human culture) produce evolution by the process of genetic assimilation. Then culturgen assimilation occurs, as the newly evolved ERs permit the elaboration of culturgens within the new limits.

However, it is important to note the possibility of a reverse process occurring. Lumsden and Wilson argue that if human beings devise new culturgens which persistently lower genetic fitness, then the process of genetic assimilation will operate so as produce ERs which are less permissive. That is, they will tend to filter out the non-advantageous culturgens – the 'leash' will tighten.

The authors consider another possible way in which maladaptiveness might be supposed to arise. This involves the hypothesis of a cultural formation which is maladaptive in genetic terms, but which has become so powerful, and able to use the tendency to conformism alluded to earlier, that it becomes autonomous of human choices, and acquires something like an independent existence. This is rather like the way that Marx envisaged the alienating character of the capitalist economic system. However, they argue that

no cultural juggernaut will persist indefinitely under such ill-fitting conditions ... If
epigenetic rules are contravened they can be expected to exert a steady pressure until the
culture is realigned into a more congenial form. (Lumsden and Wilson 1981: p. 179)

Clearly, in the case of each of these hypothetical modes of maladaptiveness
of culturgens there will be time-lags involved, which can explain why culturgens
which are not adaptive can exist for a period within a human population before
they eventually die out. This is an important point to bear in mind as a possible
response to those who argue that evolution cannot plausibly be applied to the
understanding of human cultures because there are examples of non-adaptive
behaviour persisting within human populations.

We have seen that the idea of the culturgen, necessary to the co-eveolutionary
hypothesis, seems to play a role analogous to that of the species in standard
evolutionary theory. Lumsden and Wilson extend this analogy in an illuminating
way by using the idea of the 'biogeography of the mind' to explore the ways
in which culturgens may be conceived to colonise minds. Biogeography is the
scientific study of the distribution of organisms across different geographical
areas. The particular form of the discipline which the authors find useful for the
elucidation of co-evolution is known as island biogeography, which studies how
species come to occupy those discrete geographical areas known as islands – a
term which covers any ecologically bounded area, not simply land surrounded by
water. This is the field of enquiry in which Wilson made his name in the 1960s (see
The Theory of Island Biogeography, co-authored with R.H. MacArthur). Island
biography seeks to ascertain such matters as the rates of colonisation of initially
uninhabited islands by different species, when those islands are of different sizes
and distances from the sources of those species.

The analogy which Lumsden and Wilson draw is between the individual
human mind and an island, with culturgens then being thought of as potential
colonisers of minds, just as organisms are potential colonisers of islands. This,
they accept, is a somewhat rough analogy, but suggest that it can provide a useful
way of thinking about the fate of culturgens within human minds, and thus of the
cultures which emerge from those structures Lumsden and Wilson 1981: p. 305).
Thus, analogies can be found in the case of minds for the two key features of
islands – their areas and degree of isolation from sources of species.

Corresponding to area is the capacity of long-term memory to store culturgens.
The larger the area of an island the larger the number of species it can hold in
an equilibrium state. That is, particular species come and go, but the overall
number remains constant and related to the area by a mathematical equation
called the area-species curve (Wilson 1992: pp. 208–9). Similarly, the larger the
capacity of long-term memory, and the lower the rate at which the component
neural structures decay, the larger the equilibrium number of active culturgens it
can contain. Corresponding to distance in the case of islands is the phenomenon
of relative isolation of individuals from surrounding cultures. Hence, just as the
more isolated the island, the lower the immigration rate of new species, so the

greater the cultural isolation of human individuals the lower the rate of culturgen immigration into their minds (Lumsden and Wilson 1981: p. 305).

The key idea of island biogeography is that there are limits to the numbers of species an island can contain. The reduction of an island's size must lead to a reduction of the number of species it can support. Any island, therefore, can only contain a certain proportion of the species available to colonise it. This is an important finding for the conservation of biodiversity. It means, for example, that it is impossible drastically to reduce the size of an area of rainforest, such as that of the Amazon, and expect it to retain the same numbers of species, even at reduced sizes of population.

In the case of human minds and culturgen transmission, the analogy suggests that any individual human mind will be limited in the number of culturgens it can contain. Lumsden and Wilson speculate that the number of culturgens which can be held within a single mind at a time – what they refer to as the 'packing' property – will be determined in turn by three cognitive capacities. These are the ability to discriminate between different culturgens within specific categories, the ability of long-term memory to categorise and recall culturgens, and the valuation of the stimuli associated with each culturgen.

This limitation of the mind to absorb culturgens sets up competition between culturgens for admission to human minds, and the competition may be expected to lead to the extinction of some culturgens within those minds. However, Lumsden and Wilson recognise some key differences between the culturgen and species cases. One is that the existence of societies as organised entities, existing outside of any individual minds, in itself tends to help to retain certain culturgens within the minds of the constituent members of the societies. Whether or not some culturgens successfully invade minds and oust others, therefore, is not to be explained solely on the basis of the properties of those culturgens.

A further key element operating in the case of the culturgens is that they are capable of forming linkages with each other in such a way as to restrict entry to new culturgens. That is, culturgens are not isolated atoms, they can form assemblages which protect the component elements from attack by invaders. It is not impossible to find such alliances within the world of organisms too, as when certain species of ant protect specific kinds of plant from destruction by other insects, receiving benefits in the form of food and other materials from the plant.

Examples of such assemblages in the case of culturgens are those assemblages that go by the name of ideologies, or worldviews. An attack on one component of such an ideology, such as a certain factual claim that is hard to support on the basis of evidence, may, for example, be protected by the coexistence within the ideology of a culturgen which elevates intuition above reason. Such assemblages can eventually be broken up, Lumsden and Wilson note, in the course of what we refer to as conversions or revitalisations. A new set of culturgens may then enter the mind of the human individual undergoing such an experience, and the end result may be an increase or a decrease in the share of available culturgens present in the mind in question (Lumsden and Wilson 1981: p. 305).

Lumsden and Wilson are undoubtedly right to seek to integrate these complications concerning culturgens into their account. But they do highlight an issue which has hitherto not been directly addressed, which is whether certain processes that have been privileged in most traditional accounts of epistemology, such as logic and the assessment of evidence, should themselves simply be regarded as culturgens, that may come and go within human minds as do any other culturgens, or whether they should be given a more privileged status.

If the former, then the position of Lumsden and Wilson is akin to that often attributed to postmodernism, which holds that there are no privileged conceptions which have universal validity, and which might give us a fixed point (a centre or foundation) from which to judge and perhaps eradicate certain culturgens as without merit, or even as harmful. If the latter, then the only place within the theory of gene-culture co-evolution where we can locate such processes is in the box marked 'epigenetic rules'. Doing this, however, in effect gives us a purely naturalistic account of logic and evidence assessment. That is because ERs ultimately are gene-based, and so come into existence as adaptations, produced by natural selection. Hence, on this view, reason and logic have no ultimate or transcendental status. They only feel to us this way because we have evolved to possess the ERs that embed them in our minds, and ultimately this is because those minds that possess them will better enable their possessors to pass on their genes to subsequent generations.

This implies, in turn, that we should distinguish between what we are bound to take as given or foundational, and what should be regarded as absolute, or in some way unquestionable. We have so evolved that the ERs that constitute our reasoning capacities have to be treated by us as given, but the naturalism of evolutionary theory implies that if the world had been different then we would have evolved different ERs. Nothing in evolutionary theory itself can tell us if there are any limits to what the world can be like, and thus what the reasoning capacities might be which successful interaction with different worlds might require. We might try to work out, with our present apparatus of ERs, whether or not there are any limits to possible worlds and thus whether at some level the contents of the ERs which specify successful forms of reasoning can be delineated for all possible worlds. But there perhaps must always be a suspicion that any limits that we claim to detect are thrown up by the specific ERs that we happen to possess.

For many, the idea that reason and logic can have such a contingent basis will be an intolerable one. As we will see, there will be similar qualms about the naturalistic accounts of morality. We will have to examine these issues further when we examine the implications of naturalism more directly at a later stage.

Returning to the biogeographical analogy for the mind, Lumsden and Wilson offer the final suggestion that a human society as a whole can be thought of as resembling an archipelago. Such mind-island archipelagos resemble geographical archipelagos in a crucial respect. This is that the constituent islands of the former exchange culturgens with each other more frequently than they do with locations outside the archipelago – other cultures. This is also the case with respect to real archipelagos and the exchange of species between their constituent islands (Lumsden and Wilson 1981: p. 306).

In the final part of their discussion, Lumsden and Wilson consider how co-evolution theory might be integrated into the social sciences. We will defer further consideration of their arguments under this heading until we consider the whole idea of consilience. However, it will be useful to conclude this outline of their version of co-evolution theory by noting that they explicitly deny that their theory involves any reduction of the study of human beings to the studies of other animal species:

> The learning of language, reification, and disjunctive concept formation are biological in the sense of being subject to genetically underwritten epigenetic rules, but as far as we know these processes are distinctively human. Such learning can be analyzed by gene-culture theory, but not by direct comparison with animal species. The inseverable linkage between genes and culture does not also chain mankind to an animal level. (Lumsden and Wilson 1981: pp. 344–5)

The version of co-evolution theory offered by Lumsden and Wilson contains much that is speculative, and does not flinch from seeking to find as close a connection as possible between biology and culture while nevertheless acknowledging, and trying to do justice to, the variety, changeability and centrality of human culture. They give good reason to suppose that a commitment to the Darwinian worldview does not require the jettisoning of any element widely regarded as crucial to an understanding of the human mind. Whether such a worldview can offer a fully adequate account of such phenomena is a different matter. But there are varieties of the Darwinian account available, some with a less prominent position for the biological within human culture than Lumsden and Wilson provide. Let us now turn to consider some of these other varieties of Darwinian thought.

Other Versions of Co-evolution

In spite of the considerable efforts of Lumsden and Wilson in the work we have been considering, the most influential versions of gene-culture co-evolution have developed from the work of Feldman and Cavalli-Sforza and the anthropologists Robert Boyd and Peter Richerson who published another work of co-evolution theory – *Culture and the Evolutionary Process* – in 1985. As Laland and Brown explain, there has now emerged from these alternative approaches some consensus concerning the best way to analyse gene-culture interactions (Laland and Brown 2002: p. 246). As we will note, however, many of the particular claims made by Lumsden and Wilson survive into this consensus version of the theory.

This consensus emphasises that human culture is largely autonomous of genetic influence, and thus it is a position which aligns more closely with what most social scientists are strongly inclined to believe than is common among others in the sociobiological family. Cultures change more quickly sometimes than can be accounted for by gene-based processes, and human populations living in similar natural environments can be shown to adopt very different cultural traditions. (Laland and Brown 2002: pp. 248–9). Thus, even if there is some influence upon human culture that comes from genes and the natural surroundings in which human beings live, their cultural traditions seem to have a very strong effect. Cultural traits in a given population will thus largely have come about through social learning, which means that such traits can also change rapidly in response to a variety of factors. Co-evolutionists also emphasise that cultures modify the selection pressures on genes and strongly effect the development of the behavioural traits of individuals (Laland and Brown 2002: p. 249).

Co-evolutionists aim to give a detailed account of how cultures can evolve, in terms of cultural selection. Standard natural selection is allowed to effect cultural selection via the differential survival of individuals who express different cultural preferences, as Lumsden and Wilson suggest. But behaviour that reduces genetic fitness, by leading those who adopt it to leave fewer offspring, may survive because it has high cultural fitness – spreading extensively through a population for primarily cultural reasons, such as status enhancement. Individuals have a role to play in cultural selection, by altering the culturally transmitted behaviour patterns in the light of their own experience, and then passing on to others the altered trait. This may in time produce a more exact fit with environmental factors for a given population, in the manner traced by human behavioural ecologists (Laland and Brown 2002: p. 251).

The effect of genes on culture is still allowed for in the co-evolutionists' perspective. What makes an individual choose to adopt one behaviour rather than another available one may be the result, at least in part, of a genetically-based predisposition akin to the epigenetic rules hypothesised by Wilson and Lumsden. The term 'phenogenotype' was coined by Feldman and Cavalli-Sforza to refer to the phenomenon of the non-random combination of genes and cultural elements (Laland and Brown 2002: pp. 254–5). But there are other important factors that affect the likelihood that an individual will adopt one form of behaviour rather than another. One is the sheer frequency of the behaviour in the culture inhabited by the individual. A tendency to be conformist may operate in such cases. Also, Individuals may link behavioural traits, so that the frequency of one trait may encourage that of another trait with which it is commonly linked, such as wealth and conspicuous consumption (Laland and Brown 2002: p. 252). As we have noted, versions of all these points were put forward by Lumsden and Wilson.

Another important dimension to cultural evolution concerns the pathway by which cultural elements are passed between individuals. Genes go only from parents to offspring. But cultural forms may pass not only from parents to children, but from one generation to the next, or between members of the same generation. This will mean that the outcome of cultural evolution, even when it involves co-evolution, cannot be expected to exhibit the same outcomes as genetic evolution *simpliciter* (Laland and Brown 2002: p. 253). It is possible to produce mathematically precise models of how the frequencies of genes and cultural traits will change over time within a population as the result of the interaction of natural and cultural selection and the effects of biases such as is represented by the phenogenotype phenomenon. Laland and Brown see this feature of the co-evolutionists' paradigm as permitting one useful way of dealing with the problem that so little is known about the actual course of human evolution – a problem for the sociobiological family in general. For the co-evolutionists' mathematical models can enable the feasibility of various evolutionary hypotheses to be tested, and can point towards promising lines of empirical research (Laland and Brown 2002: p. 259).

An important implication of the co-evolutionary approach is that it can produce a viable model of group selection, provided that we restrict ourselves to the consideration of cultural, rather than natural, evolution. For a cultural trait that became prevalent among a certain group, such as a certain economic practice, might enable that group to survive better than competitor groups. As its population rose, provided the cultural trait was passed on down the generations, the cultural trait would spread with the spread of the population that carried it.This is a form of evolutionary theory that has had its supporters outside the ranks of the professional co-evolutionists – as in the evolutionary speculations of Hayek, for example (Hayek 1960: p. 59).

Laland and Brown suggest various advantages to the group selection hypothesis:

(1) Group selection requires conformity to a general practice within the group and such conformity keeps the group separate from other groups: non-conformists are discriminated against.
(2) Other groups may accept the trait rather than continue to lose out, hence it can spread more rapidly than occurs in genetic evolution.
(3) Cultures contain devices, such as symbol systems, which enable them to resist casual influxes of cultural forms from other cultures; by contrast, local gene pools (demes) find it harder to resist gene flow.
(4) Culturally created forms of information about free-riders with respect to the behavioural trait, coupled with culturally-sanctioned forms of punishment for them, keep the group intact with respect to the trait in question, and strengthen possibility of group selection. (Laland and Brown 2002: pp. 264–5)

In other words, the group selection hypothesis applied to cultural evolution helps to explain some commonly observed features of human groups, such as conformity, exclusion of some traits from other cultures as alien, as well as the absorption of such cultural traits on other occasions, when the very survival of the group is at issue.

Laland and Brown note another possible implication of the group selection hypothesis put forward by Richerson and Boyd (1998), namely that if group selection has applied over a prolonged period to human cultural evolution in the manner described, then it would have created a stable enough social environment for the selection of genetic predispositions towards 'tribal instincts'. These involve behaving altruistically towards fellow group members, but with hostility towards outsiders (Laland and Brown 2002: p. 266). However, it should be noted that insofar as this hypothesis involves noting what is usually viewed as a dismal aspect of human behaviour, it is the cultural rather than the genetic that is at the heart of the phenomenon. This should caution us against the view that only those who emphasise the importance of the genetic basis for behaviour take a dim view of human nature.

As we have seen, co-evolutionists of all stripes treat culture as comparable with genes in the respect that it can be analysed into discrete elements – the memes or culturgens. We noted in the discussion of Lumsden and Wilson's version that this type of analysis is not unique to such theories, but is resorted to in some of the social sciences, such as archaeology. However, that point, even though it shows that social science cannot afford to be overly critical of co-evolution theory on this basis, does not in itself form a justification for this mode of theorising. What else might be said in its defence? Even if we adopt the view that culture is information, a view which is currently said by Laland and Brown to be in the ascendant (2002: p. 272), can it really be digitalised? Or must it be treated holistically?

The problem with holism is that it leads to enormous difficulties of analysis. It becomes very difficult to know whether one has any grasp of the whole at all, or whether any synoptic picture one may provide of the whole for explanatory purposes is accurate or comprehensive. The danger is that one is forced to operate at a level of generalisation and vagueness that does little to advance any explanatory agenda.

Laland and Brown offer a further argument, this time drawn from the natural sciences, such as biology. These, they argue, have made their most significant advances when they have sought out sub-elements within complex wholes, such as the human brain, and sought to analyse them as quasi discrete elements, even when the fact of their interconnection with other elements is fully recognised. In any case, there is compelling evidence from within psychology that human beings treat cultural phenomena in this digitalised way, understanding them and retaining them in discrete packages within the brain. When things go wrong with the brain this is often revealed in highly specific malfunctions with respect to discrete semantic elements in the individual's intellectual repertoire (Laland and Brown 2002: pp. 272–5).

On this basis a reasonable case can be made for treating the processes of cultural and biological evolution as analogous. As Laland and Brown put the point, 'both genes and memes are informational entities that are differentially transmitted as coherent functional units and that exert an influence on the phenotype' (Laland and Brown 2002: p. 275). But even if they are not analogous, Laland and Brown argue, this does not invalidate the importance of the co-evolutionist view. Those perspectives which take the view that genes and culture interact and coevolve are more likely be fruitful for human self-understanding than those which concentrate exclusively on one or the other of these elements (Laland and Brown 2002: p. 276).

Nor should it be supposed that the interactionist hypothesis cannot be successful because genetic evolution is always slow and cultural evolution is always fast. In fact the reverse is equally possible. Thus, there is nothing in the variability of rates of evolution within the genetic and the cultural that precludes gene-culture co-evolution. However, in an aside which has importance for the environmental ethics argument we will be discussing later, Laland and Brown note the possibility, and by now the reality, that cultural evolution may be so fast that genetic evolution cannot respond to it (Laland and Brown 2002: p. 279). This may be of the limited importance which they take note of – that only other cultural elements can change quickly enough to interact with any given cultural change when a certain high rate of change is experienced, and thus cultural evolution becomes the most important factor in evolution. But it arguably has some dire consequences for those life-forms which have only genetic evolution to enable them to respond to rapid changes in the environment produced by human cultural change. If they cannot change fast enough they may be heading for extinction.

Conclusion

What, finally, can we conclude about gene-culture co-evolution as perhaps the most promising, because most empirically adequate and potentially fruitful, of the versions of sociobiology currently available? Clearly, it offers the Western social science tradition a view of the relations between the biological and the cultural which many culturalists ought to find congenial. It allows for a large measure of autonomy for the cultural dimension of human life, and a large measure of flexibility with respect to the acceptance and development of culturgens by human beings. We have noted that even the version pioneered by Lumsden and Wilson allows for just such flexibility, on the basis that human euculture could not be entirely genetically programmed, given what we know about the human brain, and thus that euculture can only exist with the human brain to carry it if culture plays a large part in determining human development.

We have also seen that this version of sociobiology, even in the more biologised version of Lumsden and Wilson, explicitly excludes the kind of greedy reductionism that critics often take to be inherent in it. Human culture, although growing out of the kinds of protoculture to be found in other animal species, is so different from them that it cannot fruitfully be analysed on the basis of a direct comparison with them. Human euculture is unique among animal species, and can only be understood by means of concepts and theories designed specifically for the purpose (Lumsden and Wilson 1981: pp. 344–5).

However, even at its most accommodating to the culturalists' position, gene-culture co-evolution retains the key idea that biology and culture are linked, and that culture can only fully be understood if this fact is fully taken into account.

Even though proculture cannot be used fruitfully to understand euculture, euculture has grown out of proculture by processes that only Darwin's theory of evolution can enable us to understand. Understanding culture requires the understanding of the history of culture, and far enough back in that historical account we will need an explanation of the emergence of human euculture.

But apart from this general Darwinian point, it needs also to be understood that gene-culture co-evolution is a very new approach to the application of evolutionary ideas to human beings. Although the more biologically emphatic approach of Lumsden and Wilson seems not to be in favour amongst current practitioners of the approach, it remains possible that genes hold culture on a shorter leash than seems to be the case at the moment.

In any case, all forms of gene-culture evolution seem to give support to the idea that the social sciences, and the humanities too, should be seen as connected in important ways with the psychological and biological levels of humanity. Wilson has done much to explore this idea under the heading 'consilience'. It is to his arguments on this topic that we turn in the next chapter, before we then consider arguments based on the idea of humanism for rejecting such connections, and in particular for disconnecting an environmental ethic from any connection with sociobiology.

Chapter 5

Consilience

As we have seen in the preceding chapters, when it comes to the application of Darwinism to human beings, there is disagreement amongst Darwinists over just how much of human individual and social behaviour can be explained by reference to gene-based structures, and over how important are the structures and patterns of behaviour which can be so explained.

This disagreement amongst what one might call 'professional' Darwinists – biologists of various kinds – is also beginning to manifest itself amongst what one might call the non-professional Darwinists, such as social scientists. One may call the latter 'non-professional' Darwinists because many social scientists are happy to accept that Darwin's ideas are now an indispensable part of biology and thus have some application to the human animal. However, many are highly sceptical of the claim that human beings' behaviour can properly be explained in any very extensive way in terms of gene-based structures, in the manner of sociobiology (see, for example, Benton 1991). Nevertheless, the existence of Darwinian perspectives is gradually attracting notice among social scientists. The issue of how useful and explanatory is the Darwinian approach to human behaviour is clearly one which now needs to be addressed directly by social scientists, for, as the preceding chapters have suggested, the natural sciences are now finally in shape to begin addressing the issues held to be the sole preserve of social science for over a century.

A key implication of the sociobiological project is that what Wilson has dubbed 'consilience' becomes an important issue (Wilson 1998). According to Wilson, and to other Darwinists, such as Dennett, the natural sciences have become consilient already, and they both expect that it is only a matter of time before the natural and social sciences become consilient. Consilience means a form of reductionism obtains – though not eliminative, or what Dennett usefully labels 'greedy', reductionism (Dennett 1995: p. 82). Consilience as applied to the natural sciences means that they can be arranged in an explanatory hierarchy – physics, chemistry and biology. In Dennett's terminology, science proceeds (at least in terms of the logic of explanation, if not in terms of the logic of discovery) by constructing 'cranes' – mechanisms for erecting new structures – on the basis of what is known at fundamental levels, to build theories at higher levels. What this precludes is 'skyhooks' – theories completely disconnected from established knowledge which are lowered out of the clear blue sky in some mysterious way, like a *deus ex machina*, to do the job of explanation (Dennett 1995: pp. 73–80).

Thus, the laws of physics explain how atoms and molecules are formed. Chemistry investigates the molecules and discovers laws governing their action at the molecular level. Although chemistry does not reduce to physics, the laws of chemistry need to be compatible with those of physics. A theory in chemistry that clashes with a theory of physics creates a crisis – one or other must be rejected. It need not be the chemical theory that is rejected, however, for although physics deals with the most fundamental levels of the universe this does not mean that theories in physics automatically trump theories at the higher levels. More fundamental theories are not necessarily more certain, or well established, than theories developed to explain phenomena at higher levels.

One of the most famous examples of this phenomenon concerns the mismatch at the end of the nineteenth century between the estimated age of the Earth as calculated by the greatest physicists, such as Lord Kelvin, and the evidence from palaeontology, in the form of newly discovered fossil beds, of aeons of biological development. The latter could not be accommodated within the former. Even though physics is more fundamental in the explanatory scheme of natural science than palaeontology, in the end it was the physicists' estimates of the age of the Earth that had to be given up. The palaeontologists turned out to correct, and the planet was revealed to be immensely older than Kelvin's estimate of 20 million years. The physicists, of course, had to find new evidence from physics itself, and did so in the form of Rutherford's investigations of radioactive decay, to recalculate the Earth's age (Bryson 2003: pp. 127–9, 145–7). At that point consilience between physics and palaeontology began to be restored. The key point to repeat here is that natural science cannot simply accept such discords, or disconnections, between explanatory levels. It cannot simply brush the matter aside as representing a welcome diversity of theoretical perspectives.

Wilson has argued that biology is becoming consilient with psychology and the behavioural studies such as ethology (Wilson 1998: pp. 137–80). That is, psychology and ethology will discover laws and formulate theories unique to the levels of organisation they study, but their laws must not be in contradiction to the laws at the lower level. As before, it may be that the reconciliation is effected by a change at the lower level, but it may be that we have no option in a given instance but to reject a hypothesis at the level of psychology because it cannot be reconciled with a well-established biological theory.

Finally, what Darwinists of the sociobiological persuasion argue is that psychology will have to brought into consilience with the social sciences. Though there will be social scientific theories and explanations that operate exclusively at the level of society, they need to be tested against the lower-level claims of psychology and biology. A theory in social science that contradicts a well-established psychological theory creates a crisis. One or other will have to give – not necessarily the social scientific theory, but that cannot be ruled out either.

Here, then, is an important challenge for social scientists – to respond to the claim that the theories and explanations offered within their disciplines must be made to tally with those offered by disciplines outside their level of organisation, in the clear recognition that sometimes this will require the abandonment of some, perhaps cherished and prestigious, social scientific theory.

Wilson has conceded that the belief that consilience can be established between the natural and the social sciences is not itself something that can be said to have been vindicated by any science. Instead, he accepts that:

> It is a metaphysical world view, and a minority one at that, shared only by a few scientists and philosophers. It cannot be proved with logic from first principles or grounded in any definitive set of empirical tests ... Its best support is no more than an extrapolation of the consistent past success of the natural sciences. (Wilson 1998: p. 7)

Why should one give one's support to such a metaphysical world view? Interestingly, he does not immediately turn to Darwinism for supporting arguments, although one might have urged the Darwinian perspective at this point, and claimed that if human beings have evolved by natural selection, then there is a prima facie plausibility to the idea that scientific understanding of them must be continuous with understanding of the other natural phenomena from which they have emerged. Instead, he makes the rather weaker point that '[g]iven that human action comprises events of physical causation, why should the social sciences and humanities be impervious to consilience with the natural sciences?' (Wilson 1998: p. 9).

The claim to which he gives most prominence in making the case for the metaphysical world view of consilience is rather that it furnishes the 'prospect of intellectual adventure' and 'the value of understanding the human condition with a higher degree of certainty' (Wilson 1998: p. 7). This latter idea is an important theme throughout Wilson's subsequent discussion, especially when he comes to characterise the achievements of the social sciences in general. The main charge to be held against them in the absence of consilience with biology and the other natural sciences, in spite of the knowledge of society which they have undoubtedly accumulated, is that very few of their major claims can be said to have been established with any degree of certainty.

It is important to notice an implication of this point which may lead to misunderstanding of the case for consilience. For it is part of that case that important social scientific conclusions which lack certainty in the absence of consilience may become certain once it is achieved. In other words, the case for consilience between the natural and social sciences does not rely exclusively on its making possible new findings, it rather may achieve important outcomes by putting existing social scientific findings on a new, more certain basis. Some may not see this as a particularly important implication, for if no new knowledge is forthcoming then the attempt to achieve consilience may not seem worth the effort. However, this is to fail to notice the difference in the levels at which knowledge may be established. Knowing something is one level. Knowing that one knows it is a second, meta level. What consilience holds out as a possibility is new meta-level knowledge – knowing that we know something, that a claim to knowledge really is just that. And of course if it is really knowledge then it can be built upon and related to other justified knowledge claims. New knowledge claims will undoubtedly emerge from this.

An example at this point may serve to illustrate the claim being made here. Let us take a specific phenomenon – social stratification – which is central to political science and political sociology, and upon which many important theorists within those disciplines have had much to say. It is a topic upon which evolutionary psychology has begun to venture some interesting hypotheses. This will serve to indicate how a Darwinian approach to a politically important characteristic of most human societies may provide good reason for accepting some of the hypotheses of political science concerning that characteristic and for rejecting others, in the manner of disciplines which are consilient, or seeking consilience.

The anthropologist, Jerome Barkow, approaching the study of stratification form the Darwinian perspective, begins by remarking that the division of society into classes and castes, unequal in 'rank, wealth and power' (Barkow 1992: p. 631), requires the existence of enough of a surplus to support 'some degree of political and economic specialisation' (Barkow 1992: p. 631). Such societies are unlikely to have existed to any appreciable extent prior to the arrival of the practice of animal and plant domestication in the Neolithic period, around 10,000–12,000 years ago. Hence, stratification is quite a recent social phenomenon.

Barkow then makes two important argumentative moves. First, he assumes that social stratification could not have been the product of natural selection directly, given that it is so recent. However, '[t]he psychological traits that enable individuals to generate stratification ... presumably are the products of natural selection' (Barkow 1992: p. 632). Secondly, he assumes that such selection would take individuals as the unit of selection. He does not expressly rule out group selection, but claims that taking the individual as the unit is the 'more profitable' approach (Barkow 1992: p. 636, fn4).

He recognises the possibility that he is guilty of 'panselectionism' – the assumption that all important traits in an organism are there because they are adaptations in the Darwinian sense – but supports the adaptationist presupposition because it leads to testable hypotheses, whereas the non-adaptationist perspective does not (Barkow 1992: p. 636, fn 4). Hence, Barkow, in the course of the argument which he goes on to construct, can not properly be accused of ignorance of the important kinds of dispute amongst Darwinians which we noted earlier.

He then puts forward three 'hominid psychological traits' (Barkow 1992: p. 632) that may lead to social stratification, in certain situations. Firstly human beings, like other primates, seek high social standing. This idea, he claims, is virtually taken for granted in 'most of the social and behavioural science literature' (Barkow 1992: p. 632). Secondly, human beings everywhere 'favor their own offspring over nonoffspring, close relatives over distant, and kin in general over nonkin' (Barkow 1992: p. 633) – in other words, engage in nepotism. It will be recollected that this is a phenomenon for which Hamilton (1964) has elucidated an evolutionary account in his discussion of kin selection. Nepotistic behaviour is still prevalent even within societies for which the term has become purely pejorative. In many human societies it is still treated as a valid norm. It is also detectable within other species (he cites Krebs and Davies (1987)).

Finally, the ability of human beings to engage in social exchange, reciprocal altruism, quid pro quos, and so on – an ability for which there is an already well-developed evolutionary account (he cites Trivers (1971), Axelrod and Hamilton (1981)) – enables social alliances to form beyond close kin. This enables individuals to unite with other, non-kin, individuals to form what often amount to conspiracies to maintain the advantages that they have managed to secure for themselves and their kin.

Given these three traits, for which quite compelling evolutionary accounts can be given, and which, in respect of the first two, are present in other species, including other primates, a hypothesis can be generated which explains the formation of elites and thus of social stratification.

Firstly, those individuals who have attained the high social standing which all human beings have evolved to pursue, enter into relations of social exchange with those others able to help them sustain that high social standing. Given suitable cultural and environmental conditions, in which the surpluses spoken of earlier are generated, high social standing will involve control over those surpluses. Human beings of whom this is true will, as nepotistic beings, use this control to favour their offspring and kin. It also means that social exchange between the members of the socially-dominant group enables individuals and their families to merge into a fairly cohesive and self-interested class, with its members doing favours for the offspring of other members in return for similar favours by those others with respect to their own kin. Non-members will be excluded from the upper class group, and that class will seek to safeguard its advantages by erecting barriers to social mobility. As Barkow concludes:

> Thus, whenever a society achieves a relationship among its population density, environment, and technology such that surplus production of food and other goods reliably results, the psychology of our species makes it very likely that social inequality and, eventually, social stratification will soon follow. (Barkow 1992: p. 634)

Barkow finishes this sketch of a Darwinian account of the origins of social stratification by seeking to establish the falsifiability, and thus the scientific status, of the hypotheses involved in the account – namely, he claims that:

1 human beings strive for higher relative standing, and do so usually by trying to control the surpluses of their society and the means to generate them;
2 human beings are universally nepotistic;
3 social exchange algorithms (as specified by Cosmides and Tooby) are correct;

4   across societies and historical periods, when societies generate surpluses this
    phenomenon always involves social rank differentiation and in the majority
    of cases also involves social stratification.

The Darwinian account fails, he argues, if any of (1)–(4) is falsified.

For the purposes of this discussion, the question of the truth or falsity of this
Darwinian account is not the main issue. Rather, the importance of the account
is that, if it can be sustained, then the analyses of social stratification given by
political scientists and sociologists will have to be compatible with it. Any account
of social stratification that sees it as a purely cultural phenomenon, not based in
any universal trait of human psychology, will have to be rejected.

Further, any normative or prescriptive account of human society that
postulates the desirability of a non-stratified society will have to provide an
account of how the human traits of status-seeking and nepotism can be prevented
from issuing in the creation of elites. We thus have to elucidate a specific,
scientifically grounded program of social reform. Some favoured nostrums, such
as raising children away from parents, educational approaches and propaganda,
will probably have to abandoned as psychologically unrealistic. The creation of
institutional safeguards against nepotism might be a more promising route. But
political history reveals the difficulties that lie in that direction.

At any rate, the prescriptive task becomes somewhat clearer, as do the lines
of analysis for the further investigation of social stratification. In this way the
consilience of political science with what may turn out to be a well-grounded
Darwinian approach to human psychology promises to anchor the former in
a way which illuminates political problems and phenomena and enables us at
last to make a non-arbitrary choice between theoretical approaches within the
discipline of politics.

In the absence of such consilience, however, the social sciences are, Wilson
argues, in a parlous state. What we have within them is simply an open-ended set
of approaches to the study of social phenomena, a plethora of theories which do
not connect with each other, let alone with theories outside the social sciences,
and a series of waves of intellectual fashions which sweep through the world of
the social sciences without much justification for their temporary dominance in
the state of knowledge within the fields of enquiry concerned.

Let us consider some of the more specific characterisations that Wilson offers
of the current state of the social sciences. He claims that social scientists by and
large are strongly inclined to reject the idea of hierachical knowledge structures
which are the conceptual underpinning of the whole idea of consilience (Wilson
1998: p. 201). Within the social sciences different disciplines simply exist side by
side, paying scant attention to each other, internally divided by different theoretical
postulates, and lacking any basic set of unifying concepts or vocabulary.

No serious attempt is made by social scientists to link up their theories to
knowledge in the natural, and specifically biological, sciences (Wilson 1998:
p. 202). The theory of psychology that most social scientists are happy to use

is 'folk psychology' (which we have already noted that Dupré sought to defend against the views of evolutionary psychologists). As we noted in Chapter 2, moral and more generally ideological, considerations pervade the social sciences' rejection of scientific findings from biology and psychology – especially the fear of Social Darwinism. Wilson notes that another important factor is the humane concern of many social scientists to accord all cultures, and the people who inhabit them, equality of respect. This is used to justify cultural relativism which then, however, makes it completely obscure why certain cultural practices should be looked at askance – colonialism, child labour, judicial torture and so forth (Wilson 1998: pp. 203–4).

The social sciences in general, and sociology in particular, follow Durkheim's prescription to view society as autonomous of individual psychology, and view the latter as in any case the product of social forces – culture. Thus, biological forces of the kind championed by sociobiology have simply no scope to operate within this picture (Wilson 1998: p. 208). The preferred method of sociology is hermeneutic, involving the largely descriptive tracing of meanings and cultural formations by scholars who are allowed by the discipline simply to adopt differing culturally defined viewpoints. But no attempt is made to trace causal patterns across the biological, individual and social domains (Wilson 1998: p. 209).

Economics, Wilson agues, is the social science that has the best prospects of facilitating the beginnings of consilience with the natural sciences (Wilson 1998: p. 216). However, its worthy attempt to find the laws governing all possible economic arrangements pays scant regard to the biologically-formed basis of human behaviour, a basis which means that only a small number of such arrangements are feasible for human beings. This severing of economic models from the biologically based complexities of the behaviour of actual human beings results in their ongoing predictive shortcomings. The prestige of the subject derives not from its successes in predicting and controlling economic behaviour for the common good, but from the fact that government and business have no alternative sources of enlightenment (Wilson 1998: pp. 218–9). Economists do show some of the virtues of good scientists. They seek parsimonious starting points and generality of conclusions, seeking to explain as much as possible on the basis of the fewest postulates. But their comprehensive failure with respect to consilience means that this effort fails. They have stripped their conception of the human being too far, ignoring the rich complexity of human behaviour as revealed in biologically-grounded psychology, and have ended up achieving at best consistency of theory, but explanatory and predictive inadequacy (Wilson 1998: pp. 219–24).

The social sciences, whether grappling with the economic, political or generally cultural activities of human beings require, Wilson argues, to complete an explanatory traverse from the genetic, thence to the brain science, on to the psychological and finally to the social levels of explanation. If social scientists attempted this traverse they would discover facts about the human brain, and so about the human mind, which ought to make a difference to their social theorising.

They would discover, for example, that the brain is designed by evolution to cope with quick decision-making in complicated situations in which there is incomplete information. It is not primarily designed to be a logico-deductive machine sifting through well-articulated alternatives in a way that maximises some desired outcome. Its preferred methods of dealing with the world involve accepting what is satisfactory now, rather than waiting for the best later, and employing various rules of thumb which have worked well enough in the past (Wilson 1998: p. 228).

Anyone who has studied or taught in the social sciences will be bound to recognise the force of many of these claims. Although I am a philosopher in terms of my postgraduate training, I studied social science as an undergraduate and have taught political science and social policy in a department of politics for two decades. The disjoint and fragmented nature of these disciplines makes for a very odd teaching experience. The expectation is that one will inform one's students of the variety of intellectual approaches to the study of politics, in all their developing complexity, in an objective way. One is required to encourage them to think of the strengths and weaknesses of each of the approaches via a reading of the chief works of the canon, whilst simultaneously keeping up to date with the latest works which have made a splash and quietly setting aside others which have dropped out of vogue without ever reaching the point of saying that they have been conclusively refuted.

It is surprising to me that students do not more often ask for some clear direction through the plethora of competing views and approaches. However, they are taught the virtues of expounding accurately the alternatives to which they are exposed, and reaching their own views, at least temporarily, and for the purpose of formulating a conclusion of an essay or exam answer. In practice a great deal of what goes on in political science approximates to a form of modern history. At least, that is the element that furnishes a solid enough foundation for the discussion of alternative interpretations without descending simply into a purely intellectual game. But one knows that the variety of theories to which one's colleagues give their several allegiances rarely come into direct contact, and certainly not to the point where it can generally be agreed that one or other has a clear advantage over its competitors.

The difficulties that these features of contemporary social science create are not simply intellectual ones of confusion, inadequacy and sterility. The social sciences are supposed to furnish the knowledge base for the formulation of sound public policy. Here the unsatisfactory nature of social science in the absence of consilience has unfortunate practical implications.

To illustrate this, let us adapt one of Wilson's own examples, from the field of environmental policy (Wilson 1998: p. 8). The conservation of the remaining rainforests, crucial to the widely accepted international public aim of preserving biodiversity, requires the intersections of (at least) biology, economics and ethics – a natural science, a social science and one of the humanities. But ethics is normally conducted without reference to any biological understanding of the foundations of ethical conduct, and so risks making pronouncements which cannot find support within the biologically-structured domain of human motivation. Economics has usually been conducted as if human beings did not inhabit a physical domain at all, let alone a complex interconnecting set of ecosystems, and rests on postulates of human motivation which again fail to have much connection with biological reality. In any case, there are different schools of economics between which there exist profound differences of presupposition, vocabulary and theoretical approach. Finally the biological knowledge crucial to the formulation of a workable and effective conservation policy is not readily available to, or understood by, individuals schooled in social sciences and humanities which have deliberately cut themselves off from all contact with the natural sciences.

It has also to be said that the absence of consilience between the social and natural sciences cuts both ways. Those who possess the requisite ecological knowledge are likely by their educational background to be innocent of economics or ethical theory, and thus to face difficulties of communication with their colleagues from the other fields when it comes to the discussion of such complex practical issues. In such circumstances policymakers will probably find it very difficult, if not impossible, to achieve anything other than the most platitudinous form of words which is the most they can hope to agree upon, leaving it to some other individuals to give practical effect to their pronouncements.

This account may be something of a caricature, but it serves to dramatise the points at issue in the consilience debate. It is not an adequate response to these difficulties to point to the existence of multi-subject forms of higher education, which undoubtedly exist. That is, it is quite common to find individuals with higher education qualifications combining various social sciences and humanities, or combining natural science with a social science or humanity (physics and philosophy, for example). These forms of education certainly make the individuals who are subject to them familiar with diverse disciplines and the ability to communicate at least to some degree with specialists in those disciplines. Such individuals maybe better placed than those who specialise in only one subject to participate in the complex discussions referred to above.

However, Wilson's case for consilience is not met simply by arguing for more multi-subject forms of higher education, although these are undoubtedly going to be indispensable prerequisites of the development of genuine consilience. What he is after can best be gauged from his description of the ever-changing situation within the already consilient natural sciences:

> Disciplinary boundaries within the natural sciences are disappearing, to be replaced by shifting hybrid domains in which consilience is implicit. These domains reach across many levels of complexity, from chemical physics and physical chemistry to molecular genetics, chemical ecology and ecological genetics. None of the new specialities is considered more than a focus of research. (Wilson 1998: p. 9)

Presumably, then, consilience between the natural and social sciences, via the bridge discipline of psychology which has itself become biologised (in the form of evolutionary psychology, gene-culture co-evolution theory, and so forth), ought in the end to produce a similar set of 'shifting hybrid domains'. What these would be has to be matter of guesswork in the current pre-consilience state. But hybrids are already available within the social sciences to a certain degree. It is quite common to find sub-disciplines such as political economy, political sociology and political psychology. There is developing within economics a variety, dubbed 'ecological economics', which seeks to unify human economic activity with the findings of scientific ecology – economics conducted within an ecosystem. The same is true of at least some of the humanities, with the discipline of history often being divided into economic history, social history, political history, environmental history and so forth. Philosophy has also, of course, long been subdivided into various specialisms which take other modes of human thought as their subject-matter – the philosophies of biology, psychology, economics, history, science, and so forth have long flourished within academia.

While these phenomena ought to quell any scepticism about the very idea of hybrid disciplines, they are still not sufficient to meet the requirements of consilience as Wilson conceives of it. That is because, as noted above, these social scientific and humanities hybrids are not united by an agreed conceptualisation of their subject-matter, an agreed set of theoretical approaches towards it and an agreed set of research methods for investigating it. An economic history written by a Marxist will differ in important ways from that written by a liberal or a feminist. And each of these can be subdivided further. Certain facts will remain common to each, but the correct description of many others will be disputed. Some facts that are important from one point of view will be ignored and downplayed by the others, and so forth.

It is this lack of fundamental agreement in the basics of the various disciplines which Wilson and other supporters of consilience believe will continue until the link is established between the social sciences and humanities and the natural sciences. As we have seen, the link is clearly supposed to help to fix the fluid, shifting conceptual world of the social sciences and provide some secure theoretical base for the creation of interconnecting theories of politics, economics, sociology, anthropology and the rest. Although there will continue to be specialisms and subspecialisms, their identity will be established, as are those within contemporary natural science, by the needs of the developing fields of research, not by the need to maintain a connection with time-honoured great thinkers or canons of received classics in the field.

One acid test of how far this was being achieved within the social sciences is to examine how far they would be prepared to jettison their history. For example, the novelist Ian McEwan reported in 2006 that his son, who had recently completed a degree in biology, had been told not to bother reading anything written in the field of genetics prior to 1997 (McEwan 2006). Given the speed of scientific advance in this field this recommendation makes perfect sense. But it certainly contrasts with the case in the social sciences and the humanities, where no writing can be absolutely discarded as worthless for the development of current understanding. When I began studying politics at university in the late 1960s, for example, the work of the contemporary economist and political theorist Friedrich Hayek was regarded as little more than an expression of nostalgic longing for the eighteenth century. Within a few years, however, his works had moved to the centre-stage of political thought, both in the real world of politics and that of academia, as part of the revival of classical liberalism, or neo-liberalism as it came to be known. He took himself to be restating the lost truths of liberalism as enunciated above all by the great thinkers of the Scottish enlightenment, such as Hume and Smith.

However, in one crucial respect it may be suggested that Wilson's own critical case lacks proper awareness of the role of ideology, and of politics, in this kind of context. It might be urged that, in common with many scientists, used to tackling specific intellectual problems and to reaching agreement in most cases with his fellow scientists, trained in the same concepts and theories as he is himself, Wilson fails to see that fundamental differences of values and worldviews are an inevitable part of public debate in any political context. He may be accused of too readily assuming that once experts have achieved consilience across their fields of study, public policymaking which makes use of their expertise will dissolve the difficulties which hitherto have appeared to be as intractable as they are inevitable.

This is charge is not obviously plausible as applied to Wilson, whose experience of the bitter debates over sociobiology, if nothing else, have alerted him to the pervasiveness of ideological thinking, including its extension right into the heart of his own discipline (Wilson 1996: pp. 330–53). However, such critics arguably do have a point. Wilson does sometimes talk as if consilience between the social sciences and the natural sciences will import into former the ready agreement on diagnosis and action that frequently characterise the latter. But once one starts to study human beings one is inevitably caught up right from the beginning in difficult and intractable value disputes, disputes that are at the heart of ideological thinking.

Consilience will, thus, inevitably come at a serious price for natural scientists such as Wilson. For they will have to abandon their professional non-involvement with values, a non-involvement which it is possible to maintain in the study of ants, for example, to take Wilson's own field of specific expertise, but which it is wholly impossible to hold on to in the study of human beings.

The argument for this point is well established. It rests on the idea that the study of human beings requires a study of human behaviour in various situations. The way in which people behave is determined by their beliefs about the situation in which they find themselves – the threats and opportunities they face – and their desires. The latter presuppose some conception of what is desirable and what is not. 'Desirable' means 'worthy of desire' and this immediately imports a value notion into the situation. To understand the behaviour one has to grasp the value judgement embodied in the desirability-characterisation.

This is turn requires the observer to have an appreciation of what a human being can intelligibly desire. Human observers of human beings thus have to make their own judgements of what is desirable – they have, qua observers, to endorse a range of possible values. They also have to make their own judgement about whether, given the range of values it is intelligible to pursue if one is a human being, the human beings one is studying are effectively identifying and pursuing the source of those values. If one judges that they are mistaken in their chosen courses of action, one will have to develop some account of how the mistake has come about – inexperience, deception, self-deception and so on are all possible explanations.

But then the matter is made complicated by the fact that the concept of values is a far from simple notion, and the more complex the being one is studying, the more complicated it becomes. One's view of what is valuable, for example, is deeply affected by one's metaphysical view. Different metaphysics provide different ways of conceptualising the same phenomena. To take an obvious case, mortality is viewed by the atheist and the religious believer in very different ways, and different religions take very different views about the significance of death and the possible forms of the afterlife. One's attitude to death, and judgement concerning the relative desirability of life and death in different circumstances, will obviously be affected by which, if any, of these rival metaphysical views one finds reason to adopt.

Perhaps metaphysics is a dispensable element in the human condition, but that seems doubtful. In any case, consilience itself has already been characterised by Wilson as, at least currently, a metaphysical position, which suggests that at least some metaphysical positions are worthy of serious consideration. Notoriously, rival metaphysical positions are hard to refute intellectually and hard to shake belief in once the commitment has been made. For each position has a tendency to reinterpret or deny the factual basis upon which their critics rest their case.

The human observers of human behaviour – the social scientists – are not exempt from any of these considerations. They too will have some view or other, however inchoate or schematic, of a metaphysical kind. This will make a difference to their interpretation of what it is reasonable to desire, and thus to their view about what is really going on in the world of human action. What social scientist A will regard as evidence of confusion or deceit (false consciousness, perhaps) in the behaviour of certain human beings social scientist B will regard as evidence of their possession of clear-sighted self-interest, or at least of an intelligible value position. The former might try to find ways of opening the eyes of the deceived to their mistake, and to enjoin a course of action, perhaps involving radical changes to their behaviour, designed to lead to a more effective attainment of goals. The latter may counsel them not to be misled by claims that they are mistaken, but keep on behaving just in the way that they have been hitherto. In other words, when it comes to enjoining action, some social scientists will be revolutionaries, some will be defenders of the status quo.

Value-free social science, for these reasons, seems to be a chimera. In the face of these difficulties it might be supposed that the only hope for value-free social science might rest with the possibility of non-human social scientists. These might possess completely different value preferences from human beings, and thus might be thought to be the only observers capable of offering a theoretical perspective exempt from these problems. But such observers would have a different problem – of making sense of human behaviour in the first place, at least above the rudimentary levels involved in simple self-preservation and reproduction which might be supposed to be common to all life-forms in the universe. Human observers of non-human life-forms on this planet, such as Wilson himself, are able to achieve satisfactory modes of theoretical understanding of those other life-forms precisely because they do not have to postulate any very rich forms of valuational activity on the part of the objects of their study.

Can the defenders of consilience offer any convincing riposte to these arguments? If they cannot, then the ideological divisions within the social sciences and humanities that are supposed to be rendered obsolete once consilience is achieved will instead have to be understood as an ineradicable part of those intellectual ventures, however strenuous is the attempt to connect them up with the natural sciences. The matter might appear to be even more intractable if it should transpire that different value positions are a reflection of different behavioural and personality traits for which an explanation in terms of epigenetic rules can be found. Then consilience may seem to be hoist by its own petard.

The key to a workable response would seem to be to try to restrict the scope of value differences within social science to those that reflect solely personality differences between people. After all, ways appear to have been found within liberal societies to allow scope for the forms of life which reflect different valuational judgements concerning life-plans deriving from the different personalities which individuals clearly have. This suggests that valuational differences emanating form this source will not be so great that they give rise to wholesale differences of worldview. The more difficult problem concerns conflicting metaphysical presuppositions and the different interpretations of 'the facts' that in turn underpin different judgements about how the individual and society ought to be structured.

This brings us to a recognition of the scope of the ambition which underpins the idea of consilience when it is brought to bear on the humanities and social sciences. It must aim for nothing less than a reduction of the variety of metaphysical positions to those which represent only a narrow range of possibilities, perhaps, in the end, to only one. If that is not achieved, then there is no good reason to suppose that such consilience as can be produced between the natural and social sciences/humanities will be sufficient to produce anything like the consilience between the different fields of enquiry as characterises the natural sciences.

Wilson does recognise that this is indeed the outcome which defenders of consilience from a Darwinian perspective are led by their position to anticipate.

In his chapter discussing morality and religion he concludes that in the end the Darwinian world view will triumph over non-Darwinian transcendental metaphysical and religious positions:

> For centuries the writ of empiricism has been spreading into the ancient domain of transcendentalist belief, slowly at the start but quickening in the scientific age ... The eventual result of the competition between the two world views, I believe, will be the secularization of the human epic and of religion itself. (Wilson 1998: pp. 295–6)

The defensibility of this view is something to which we will return in a later chapter when we consider the relations of the Darwinian worldview to religion. The important point to note at this stage is that this conquest of metaphysical transcendentalism by Darwinian consilience is only a part of the battle. There are other metaphysical positions than religious ones, and both religious and non-religious ones need to succumb to the Darwinian worldview if the defenders of consilience are to achieve their aims with respect to the social sciences and humanities. It is this latter point which Wilson does not appear fully to acknowledge.

Is Darwinian consilience up to the task? It is rather difficult at this stage to answer this question with any degree of certainty, partly because the output of the sociobiological perspective with respect to matters of direct relevance to the social sciences and humanities is still, as we noted in earlier chapters, rather piecemeal and spasmodic. If the kind of argument put forward by Barkow concerning the origins of social stratification could be fortified by extensive empirical research, and linked in a coherent way to other findings concerning key social phenomena, then we would be better placed to assess how far consilience emanating from the Darwinian perspective could hope to render unified and coherent the whole field of social science.

But the thrust of the argument is certainly a powerful one. Consilience does appear to have developed to a high degree in the natural sciences, and the latter do clearly have an undeniable tendency to interconnect and become mutually supportive. Darwinian biology does appear to have become thoroughly integrated into the corpus of the physical sciences, so much so that those who would like neatly to separate it from the rest of natural science with an eye to discarding it and leaving the rest of science intact will now find that an almost impossible task. The application of sociobiological approaches to the study of human beings does appear to be gaining ground and to be beginning to achieve useful results. The connection with the rest of an overwhelmingly successful natural science corpus will undeniably give this approach important clout in the coming years. So the momentum appears to be favouring Darwinian consilience.

From the point of view of the present discussion, which is examining the implications of the Darwinian worldview, the case for consilience as it applies to the social sciences, at any rate, does not appear to be a matter for serious alarm. If consilience develops in the manner projected by Wilson then, as he expresses it,

The social sciences will continue to split within each of its disciplines, a process already rancorously begun, with one part folding into or becoming continuous with biology, the other fusing with the humanities. Its disciplines will continue to exist, but in radically altered form. (Wilson 1998: p. 10)

This is not the same as predicting the demise of the social sciences, but is a continuation of a process of transformation with which social science has been familiar throughout its history. Granted, if it is correct it will involve a rapprochement with the natural sciences which individual representatives on either side of the current divide might not be happy with. But, given the logic of consilience, which aims to avoid greedy reductionism, there ought to be plenty of continuing distinctiveness to make the level of the social a continuing focus of distinctive intellectual endeavour.

Wilson has often been accused of greedy reductionism as a result of the kind of view expressed in the passage cited in the last paragraph and certainly the reference to social science being 'folded into' biology is hard to justify, given the logic of his position as outlined in this chapter. As we have already noted, consilience between biology, psychology and social science will make a difference, not just to social science and psychology, but to biology too. Biologists will have to acquire greater familiarity with cultural and valuational phenomena to the extent that they approach an understanding of the human brain. Sociobiology is not simply biology. Arguably Wilson has not fully enough taken on board this point, although he does show considerable knowledge of the social sciences and humanities in his whole discussion of consilience. Provided that what he writes is understood in its full complexity the charge of greedy reductionism can be rejected.

However, there are still important arguments to consider which involve the idea that in various ways the arrival of biology at the door of the social sciences and humanities poses a serious threat to much that human beings have come to hold dear. In the second part of this book we will be looking at further arguments, drawn particularly from the humanities, against the admission of Darwinism into the distinctively human domain, beginning with concerns about the threat that sociobiological approaches pose to the tradition of humanism, and then looking at the implications of such approaches for morality and religion.

PART 2
THE MORAL ISSUE

Chapter 6

Darwinian Naturalism, Environmentalism and Humanism

In Chapter 1 we noted what appeared to be a significant point of connection between much modern environmentalism and Darwinism. The former holds that human beings are part of the natural world, not set over and above it, which means that human beings remain integrated into ecological systems, local and global, even while they change them, intentionally or otherwise. This fact of ecological interconnectedness is the basis upon which environmentalists have tried to develop ethical systems which attribute to human beings various moral responsibilities towards, and with respect to, the rest of nature (see, for example, Norton 1991)

This same guiding thought has operated in the natural science of biology since the advent of Darwin's theory of evolution. Human beings are there treated as another species of organism that has come into existence as the result of the processes of natural selection. Darwin himself began this project of applying evolutionary explanation to human beings, and as we have been discovering, in recent decades various schools of sociobiology have attempted to explain the development and structure of the human brain, and thus (on the basis of materialist presuppositions which remain philosophically controversial) the human mind in evolutionary terms.

However, in spite of the shared premise, there is no obvious logical connection between sociobiology and environmentalism. Many sociobiologists have in fact had nothing to say about environmental ethics, even if they are beginning to develop accounts of the evolutionary origins of human moral thought and action. On the face of it, given the usual interpretations of the naturalistic fallacy, it would be compatible with one's acceptance of any evolutionary account of the origins of human morality that one should hold humanity to have no moral responsibilities towards the natural world.

Contrariwise, many evolutionary ethicists are positively hostile to sociobiology in any of its forms. They regard it as a specimen of the baleful rationalism and scientism bequeathed to us by the Enlightenment (see, for example, Smith 2001: pp. 121–5). These intellectual postures, in their view, have served to justify the purely instrumental and exploitative attitudes to nature that have created our current environmental difficulties. For such environmental ethicists the alternative view of human beings as having an essentially cultural mode of existence, in independence of their biological origins, is the preferred theoretical perspective.

However, some thinkers, of whom Wilson is the obvious example, have sought to unite the two endeavours – both to develop an environmental ethic, and to give their support to the project of sociobiology. As we have seen, Wilson espouses a Darwinian version of naturalism – the attempt to explain human beings and their behaviour with many of the same concepts employed to explain non-human organisms. He seeks to achieve important forms of interconnection between the social sciences and humanities on the one hand, and the natural sciences of biology and a biologised psychology on the other. The theory of evolution by natural selection is held to be the key idea – it unites the human species with all other life-forms, at least on this planet.

Hence, Wilson has written at length both in defence of naturalism and consilience and in defence of an extensive environmental ethic, particularly directed towards the preservation of biodiversity (Wilson 1998; 1992; 2002). He has sought to draw a connection between the two, finding the possibility of concern for non-human nature in an evolved tendency to value life that he has dubbed 'biophilia' (Wison 1984). In adopting this approach he has not been entirely on his own. Another well-known environmental ethicist who has found a promising concatenation of ideas in sociobiology and environmental ethics is Callicott (1989).

How might this project deal with the naturalistic fallacy argument, already noted? The naturalistic fallacy takes its stand on the impossibility of any direct logical connection between fact and value. But this leaves open the possibility that factual claims about human nature may have an indirect bearing upon moral judgements. One might develop this possibility by arguing that an empirical theory about human nature does have a relevance to the issue of how human beings ought to behave, morally speaking, precisely because morality is not just any system of norms and prescriptions. It is a system of norms and prescriptions with a specific aim – to protect the well-being and interests of all beings whose interest and well-being counts.

This leads on to the claim that the idea of interests and well-being cannot be articulated without taking up some position with respect to the issue of what kind of beings these are – what harms or benefits them; what is in their interests. An accurate grasp of their nature will be crucial to this endeavour, and it may plausibly be argued that all ethical systems take a view concerning what harms or benefits human beings and other beings worthy of moral consideration – known in contemporary discussions of environmental ethics as 'morally considerable beings' (Goodpaster 1978). Of course, as we have had plenty of occasion to note in earlier chapters, the idea of human nature is a battleground, with many of the protagonists arguing that there is no fixed human nature. But even so, the articulation of a moral system requires some view about what human beings are like – even if there is no fixity in this view, and so no possibility of a definitive system.

If this line of argument is correct, then sociobiology and environmental ethics may well turn out to have important interconnections. The view of human nature put forward by sociobiology will inevitably have implications about what conduces to human well-being. This is the point that we emphasised at the end of the last chapter, when we were considering the implications of consilience for both the natural and the social sciences. This in turn will have implications about how (at least) human beings should be treated, given the aims of a moral system. And, given that other things than human beings will have interests and the possibility of well-being, an environmental ethic directed towards, and not simply with respect to, the non-human will, prima facie, become open to development.

On the face of it, this possibility ought to make the sociobiology/environmental ethics connection at least worthy of detailed and sympathetic investigation by environmentalists. However, Lewis Hinchman has recently put forward an argument designed to persuade environmental ethicists not to pursue, and indeed actively to attack, the project of interlinking sociobiology and environmental ethics. Hinchman offers a critique of this project that focuses upon what he takes to be important moral values that apply to human beings, values championed by the tradition of humanism. He does not seek to show that the unification of sociobiology and environmental ethics will lead to an objectionable view about human obligations towards, or with respect to, the natural world (though he thinks it is very hard to produce such an ethic from the sociobiological starting point). Rather, he seeks to show that sociobiology, in its theorising about human beings, produces a view of human beings, and of moral thought in particular, which cannot sustain, and, indeed, actively undermines, key humanist values. Insofar as these values are of importance to environmental ethicists in general, the latter have good reason to break off any attempt to participate in the project of uniting sociobiology and environmental ethics.

Of course, if environmental ethicists were not to find these humanist values of interest it looks as thought this argument would not cut much ice with them. However, the values in question – self-determination and the resistance to dehumanising forms of categorisation – are ones with which most environmental ethicists will wish to be associated. This is because concern for the well-being of all species encompasses concern for the well-being of the human species, and there are powerful arguments for the view that self-determination and the resistance to dehumanising categories are central to human well-being. How, then, does Hinchman seek to show that humanist values are undermined by, or incompatible with, sociobiology?

He begins by characterising humanism throughout the ages as having a key set of adversaries – religious, scientific and bureaucratic forces which have from time to time 'seemed on the verge of reducing people to mere objects, devoid of will, dignity and choice' (Hinchman 2004: p. 4). In the course of combating these dehumanising forces, humanism has developed some key concepts which, Hinchman claims, are among its 'outstanding accomplishments, including the notions of individuality, dignity, autonomy and self-government' (Hinchman 2004: p. 4).

A key theme in Hinchman's version of humanism is the interpretavist perspective. He is a culturalist, and rejects scientism, emphasising the centrality of narrative, symbols, and hermeneutics in human self-understanding and the exercise of human autonomy. As he tells us early on, '[w]e understand ourselves, our situation, and our political alternatives in the light of narratives that we construct linking past, present and future' (Hinchman 2004: p. 4).

Developing this theme, he later goes on to argue that:

> Early humanism thus veered from the intellectual trail that would eventually culminate in the construction of mathematically based physical sciences ... Instead, it pursued a course that would lead via Vico and Herder to Dilthey's conception of the humanities as methodologically distinctive, guided by an empathetic grasp of human action 'from the inside' (*Verstehen*) and oriented to historically and geographically specific cultural forms. (Hinchman 2004: p. 8).

This concern with the unique and historical concreteness of human beings remains central to contemporary humanism, according to Hinchman. Human freedom and autonomy requires the rejection of 'mass society', a culture devoted to the immersion, and so loss, of self in 'labour, consumption and the life processes' (Hinchman 2004: p. 12). Humanism rejects, in the intellectual sphere, 'naturalism, biologism, and behaviouristic social science' (Hinchman 2004: pp. 12–13). Both of these trends threaten the loss of historical awareness, the shrinking of our consciousness to encompass only the present and the prevention of '"conversations" with the past' (Hinchman 2004: p. 14). Another area of shrinkage which humanism combats is that of the private sphere, as our lives are regimented into a set of public spaces of interchangeable significance, devoted to work and consumerism.

Thus, Hinchman's account of humanism seems to rule out the possibility that one might accept the emancipatory impulse of humanism but also view the scientific understanding of human beings as an indispensable means to achieve that emancipation – by, for example, revealing the truth about our nature, as opposed to all the fantasies and wishes with which our self-understanding is distorted.

At this point one might compare this outlook with that of Mary Midgley, which couples a defence of the evolutionary, naturalist, understanding of human beings with an excoriating attack on what she takes to be the eliminative reductionism of sociobiology – at least in some well-known versions of that approach, such as that put forward under the 'selfish gene' banner (Midgley 1994). She thus sides with Hinchman's position in attacking sociobiological 'reductionism', but unlike him strongly defends the importance of the evolutionary approach to the study of human beings, especially with respect to the phenomenon, central to human life, of morality – encompassing the key humanist values of freedom and personal integrity.

We will return later to a more direct consideration of Midgley's views. They represent an alternative conceptualisation of the connections between biology and human self-understanding which gives a larger role to the former than Hinchman appears to allow, while arguing against reductionism in a way that seems to save the personal viewpoint and human freedom which he thinks to be so important. It will also be necessary, however, to determine whether the form of reductionism which Midgley rejects can properly be laid at the door of sociobiology.

Occasionally Hinchman seems to be moving in the direction of Midgley's position, for he allows that 'mathematically based physical sciences' have their legitimate place, even within some aspects of the study of human beings. But then he maintains that they must be scrupulously excluded from human self-understanding at the level of humanity proper, on pain of producing the reduction of people to mere objects, devoid of 'will, dignity and choice' (Hinchman 2004:p. 4).

He nevertheless makes one key distinction within the natural sciences, between the 'historical natural sciences' (Hinchman 2004: p. 9) and the non-historical ones. The former comprise such sciences as biology and geology, with respect to which Hinchman makes use of Hargrove's point that these sciences also 'generated historical statements that were 'singular, contingent and historical' (Hargrove 1989: p. 78). We have, then, a distinction between 'geometric' or 'Galilean' natural science aiming to produce 'abstract, universal, laws' (Hinchman 2004: p. 9) and the 'historical natural sciences and ecology as well as the humanities'.

It is possible, Hinchman argues, to 'trace a line of development that runs from Renaissance humanism' (Hinchman 2004: p. 9) to both of the latter modes of thought. This is helpful for Hinchman's attempt to reconcile humanism and environmentalism, for environmentalism is said to be inspired by the historical natural sciences with their focus on the concrete, specific and unique. If so, then humanism and environmentalism share a common intellectual root.

However, we should note in passing that this tracing of a common root, although helpful for Hinchman's project of reconciling humanism and environmentalism, causes problems for his attack on sociobiology. This is because, since biology is here cited as a 'historical natural science', the question immediately arises as to how he can properly object to the interest shown by environmentalists in the sociobiological approach. The latter involves acceptance of the process of natural selection – aspects of which might be expressable in abstract, mathematical terms, but the actual course of which, as Gould has strongly emphasised, is probably a completely unique, contingent and unrepeatable sequence (Gould 1989). This seems to align it with the historical, unique, narrative-laden mode of human self-understanding that Hinchman makes so central to his account of humanism.

Having presented his version of humanism, Hinchman notes the similarity between the contemporary critiques of environmentalists and humanists with respect to the dehumanising tendencies of mass, consumption-oriented, disenchanted societies. He therefore, as we have noted, seeks a rapprochement between humanism and environmentalism. In order to achieve this, he argues, each mode of thought must first correct its own errors. Humanism is to give up the denigration of nature and the treatment of it as a 'sphere of heteronomy'. What must environmentalism give up? It must refuse to 'embrace a reductionist naturalism and biological determinism' in its laudable quest to 'discourage *hybris* and reintegrate people with nature' (Hinchman 2004: p. 16). Humanists and environmentalists are both concerned about the 'destruction of the life-world' – a rich, unique, tradition- and culture-saturated, experienced world within which the individual finds his/her meaning and identity. The objectifying natural sciences, reducing the world, of both human beings and non-human nature, to a lifeless mesh of timeless abstractions are implicated in this destruction (Hinchman 2004: pp. 16–17).

He needs to show, then, that naturalism is necessarily 'reductionist' in some objectionable (reifying, objectifying) sense; that biology is determinist (and thus all biological phenomena are to be explained in terms of efficient causality in accordance with universal laws). He needs also to show that treating human beings as evolved creatures involves environmentalists who do this in such pernicious forms of reductionism and determinism.

But, suppose that naturalism is not reductionist in any pernicious sense – not greedily reductionist – but requires only that social sciences/humanities be consilient with psychology/biology, which, as we have seen, means that there should be causal connectedness across levels of explanation, but not that the 'higher' levels reduce without remainder to the lower? Suppose that biology is not determinist, if by this is meant that it gives rise to predictive techniques? Suppose that biological explanation, insofar as it is evolutionary, is always irreducibly historical? Or that biology does not readily issue in universal laws – as many aver (Sterelny and Griffiths 1999: pp. 364–8)? Would it be acceptable then for environmentalists to talk about human beings and the rest of nature in a vocabulary drawn from, inter alia, the biological sciences?

Probably not. What Hinchman appears to be against is not reification, objectification, reductionism and determinism, but the replacement of hermeneutics with any other mode of explanation of human behaviour whatsoever, whether that be reifying or non-reifying, reductionist or non-reductionist, determinist or non-determinist. What he seems to support is a picture of human life as largely dependent upon stories which people tell about themselves, which have unique meaning for them, and in terms of which they find their meaning and identity. He is happy to let this story-telling approach be spread beyond human life to encompass the non-human. What he seems unwilling to countenance is anything that replaces, or even supplements, the stories.

This seems to imply that the stories are each unique, and do not embody any universal structures or forms. That is, to avoid the encroachment of scientising into narrative, structuralist approaches to narrative presumably would have to be eschewed. It would presumably be permissible for stories to make use of any universal traits in human and non-human life (such as that we are all mortal) – for these are recurring themes in all the stories which people have constructed to make sense of their world and their experience of it. But these are not the story itself. Human freedom is the freedom to make up stories about ourselves, constrained only by our own imaginations and what it takes to get others to listen and accept them as a form of 'truth'. Humanism thus seems to aim to preserve a symbolically rich, freedom conferring, endlessly changing, world of human narrative, embodying meanings and values.

Thus, on this view, although the biological sciences may be recognised to contain an irreducibly historical element, and thus not to be like the threatening, rationalist sciences such as physics and chemistry, this would not be the right kind of history. It is too 'objective', too 'given', too theory-driven. It lacks the warm, rich, yeasty quality of human narratives, their uniqueness and infinite variability. Hence, in spite of the warm words offered earlier which seem to align the historical natural sciences with the humanities, in the end the biological sciences are too objectifying and reifying in their import to be acceptable approaches to the human world, for a 'humanist'.

However, cracks arguably begin to appear in this position when Hinchman goes on to defend humanism against the environmentalist charge that it is inherently anthropocentric. He claims that 'self-determining freedom', one of the hallmarks of humanism, can only be exercised against a 'given' background of intelligibility established by institutions and cultural traditions that place demands on the moral agent (citing Taylor 1991). He says that 'the matter of such decisions ... is anchored in objective factors, *including* obligations to the natural world' (Hinchman 2004: p. 18). He uses the phrase 'objective factors' again a little later: 'humanism's commitment to freedom does not suggest that anyone should ignore the claims that objective factors such as the health of the land, wilderness or species preservation exercise on moral agents' (Hinchman 2004: p. 18).

This does not look like a completely hermeneutic picture of human life. There are objective factors, apparently, which constrain decisions – though apparently not sufficiently to remove self-determining freedom. This obviously prompts the question of how these 'objective' factors are to be identified. Is there room for natural science here – for example, in determining what is the 'health of the land'? Granted, 'health' is a value term, so that there may be expected to be irresolvable differences of opinion about what is healthy. But medical science has a lot to say about the causes of human health and sickness – and it is a natural science. If human minds can be ill, or at least not thriving, just as can human bodies, why shouldn't scientific psychology, perhaps based upon evolutionary theory, also have something pertinent and objective to say on the matter of the conditions for sustaining human psychological health?

In other words, some room appears to have been created here for the possibility that sociobiology and evolutionary psychology may play an important part in the understanding and explanation of the distinctively human part of human life. However, this possibility is supposed to be extinguished by the direct arguments that Hinchman brings to bear on the whole project of sociobiology and evolutionary psychology. Let us now turn to consider these arguments.

The cardinal sin of many environmentalists, from the point of view of a humanist, is apparently their tendency to adopt naturalism or biologism 'as the obvious conclusion from discoveries in the life science' (Hinchman 2004: p. 20). This then threatens to make ethics into a branch of (socio)biology (Wilson and Callicott are cited). He makes some immediate concessions – human beings are 'subject to nature's laws, descended from earlier primates and human intelligence, emotional life and some rudimentary kinds of human behaviour are the products of evolution' (Hinchman 2004: p. 20).

The impression given is that these concessions are not terribly important. But the latter two, at any rate, are of epoch-making importance – for they align the human species with the rest of organic life in a single system of explanation. As Midgley emphasises, we can begin to treat our attempt to understand other creatures (who seem to lack culture and a richly hermeneutic form of consciousness) as connected with, though not as identical with, our attempts to understand ourselves (Midgley 1994). Arguably this alone is enough to justify the acceptance of naturalism by those environmentalists for whom the key thought is that human beings are part of nature, not set over against it as members of a different order of being.

Hinchman goes on to make a further important point which takes him towards the position of naturalism: 'The challenge for the life sciences as well as for humanism is to investigate the evolutionary sources of human behaviour without *reducing* the internal, symbolically mediated experiences of human affairs to an external, mechanistic series of explanations' (Hinchman 2004: p. 20).

What he is against here is apparently such suggestions as the following:

1 moral obligations to the non-human world can be more firmly anchored in biological 'facts' than in cultural accounts (he cites Partridge 1984 and Callicott);
2 many symbols and products of the imagination are grounded in evolutionary history (for example, fear and awe of serpents and attraction to particular landscapes (he cites Wilson 1984)).

With respect to (1), he argues that:

> the idea that people may have evolved a moral sense with a particular content should be rejected, on the basis that not everyone shares whatever sense and specific content is proposed, and then they have to be diagnosed as psychologically abnormal. (Hinchman 2004: p. 21)

With respect to (2), he makes two main points:

a) such claims are the embodiment of an abstract, artificial, detached, view of reality. Instead of this, humanists require us to realise that in reality 'we are always already enmeshed and enthralled – as participants, partners in dialogue, speakers of language and choosers of our courses of action' (Hinchman 2004: p. 21)

b) Even if the evolutionary points are sound

> they can never completely explain our species' symbolic interactions, since these are *in toto* emergent properties, greatly overdetermined with respect to any possible biological antecedents. (Hinchman 2004:p. 21)

Then he offers a general observation:

> The move to objectivise human conduct after the manner of Callicott and Wilson also has a cost: it now becomes difficult to say why anything, human or non-human, has worth or dignity at all, if it is just part of the machinery of evolution, DNA sequences or chemical reactions. To make a case for the dignity and moral status of a human being, or of an animal species or landscape, requires not just showing that we have an instinctive attraction to it, but that such sentiment is legitimate, rationally grounded and justified, that we ought to have it. That is a question …. which biology, by its very methods, is incapable of answering. (Hinchman 2004: 21)

The following reply may be made to these claims. First, how does one show that one's sentiment is 'rationally grounded and justified' without pointing to natural facts about typical human reactions and building on these? Hinchman has already conceded that human 'emotional life' is the product of evolution. But if it is, then it must have a distinctive shape and character. This fact is arguably essential to moral debate and discussion. All moral discussion is ultimately *ad hominem*, resting on the question, concerning a proposed mode of conduct with respect to one's fellow moral beings, 'can you really live like this?'. It is the evolutionarily-produced distinctive character of the human intelligence/emotion complex which Midgley so forcefully argues to be at the heart of human moral life (Midgley 1994: pp. 128–84).

Further, all the traditions of moral argument, from Socrates onwards, posit a view of human nature, whether emotional, rational, or both – and argue for some judgements and reactions which are held to be natural or basic (even Kant thought a rational being cannot but value its own rationality). If biology and evolutionary theory can give us good reason to suppose that we do have an evolved human nature, it can help us to judge the issue between these rival accounts. This is not to say that it can settle any issues of substantive moral debate and argument – for, as Hinchman correctly notes, that always involves the judgement of many competing claims and arguments.

Further, we already do judge some people as psychologically abnormal – precisely because they do not have normal emotional reactions or think in normal ways, so that they cannot feel the force of moral argument. We think we have to guard ourselves against such people, rather than argue with them. Sociobiology may help us to clarify what we mean by 'normal' in such cases and help us to distinguish the moral eccentric from the psychopath. Of course, if sociobiology does intend to unearth genuine human moral universals with respect to the content of morality, it will count against any particular claim it offers that large numbers of human beings do not subscribe to a morality with that content. The 'abnormality' explanation risks becoming ad hoc, and so valueless, if used too readily.

What these replies show is that there is nothing new in the kind of consideration that is emerging from evolutionary perspectives. Sociobiologists, in their theorising, are not doing something new and threatening in the field of ethics which humanists must resist. They are simply contributing to very old debates on a new basis. What Hinchman is perhaps threatened by is not the 'objective, reductionist' view of human life which he claims to detect in biology, but the sense that the endless debates about what human nature is really like may finally be being answered in a definitive way.

This, as we saw in the last chapter, is a key idea behind the project of consilience. It is this definitiveness or closure that he may be seeing as the real threat to the millennia of debate. But, at least with respect to moral debate, that prospect should frighten nobody. Firstly, because, as already noted, no substantive moral issues are resolved by such knowledge, even if they are affected by it. Secondly, because human nature is not fixed by evolution – for nothing is fixed by evolution. We can envisage ways to alter our nature, if we choose to do so, although such alterations will always be based on where we currently stand and so be affected by the starting point (see Richards 2000: pp. 115–17). There is no prospect that substantive moral debate will be in any meaningful sense foreclosed by any findings in biology.

More generally, finding out the truth about human nature (the human brain) as an evolved phenomenon need not threaten anything important about human life, any more than finding out the truth about the rest of the human body did. As a result of the latter we have better medical science. Finding out the truth about the natural world demolished some options thought to be live – perpetual motion machines, and so forth. Just being able to explain in terms of efficient causality any aspect of human life and behaviour does not threaten with redundancy other levels of explanation. Finding, for example, that we have an evolutionary-produced preference for certain types of landscape tells us nothing about why one landscape artist's work is preferable to another's. Even if we may hope to achieve evolutionary explanations at even more detailed levels of aesthetic preference, individual creativity is not thereby threatened.

How should one reply to Hinchman's more general point, that 'it now becomes difficult to say why anything, human or non-human, has worth or dignity at all, if it is just part of the machinery of evolution, DNA sequences or chemical reactions'? One may first ask why should the fact that we are made entirely out of matter mean that we could not possess worth or dignity? This looks like a statement of anti-materialism. To vindicate this position Hinchman needs more than simply the idea that there are emergent properties between the biological and psychological levels. For there are emergent properties even in the levels of reality which nobody nowadays would classify as anything other than material.

It is hard to escape the view that what lies at the back of all this is old-fashioned dualism and the 'mind-first' view of reality (Dennett 1995: pp. 26–8). This holds that only mind has dignity and worth, and mind is a separate substance from matter, however complex the latter is. But naturalism does not require us to deny that the mind does encompass a set of emergent properties. As we have seen, what naturalism does require is consilience.

It is this conception of consilience which provides an alternative to the eliminative or 'greedy' reductionism which Midgley (and, probably, Hinchman) attributes to sociobiology and which leads her to reject it so forcefully. Eliminative reductionism seeks to eliminate one set of concepts and replace it with another one which is regarded as superior, perhaps in virtue of being more fundamental in some sense. But as we saw in the last chapter, the view which Wilson has propounded of the interconnections between the natural and social sciences does not seek to eliminate the latter, but to bring them into contact with the former in such a way that phenomena identified at the level of biology can be used to explain (and, admittedly, perhaps in some cases to explain away) the existence and character of phenomena at the psychological and social levels. As we noted in the last chapter, this can work both ways. An explanation at the lower levels may be rejected because it conflicts with elements already firmly established at the higher levels.

The aim is for a total, coherent system of explanation, in which phenomena at different levels are interconnected. The justification for distinguishing between 'lower' and 'higher' levels in such a picture is not that one set of explanatory theories and concepts is the only valid one, but that the pattern of explanation as a matter of fact runs mainly from the lower to the higher levels (physics explains chemistry, say, and not vice versa) and thus that if a higher level of phenomena exists at any point then so must the lower ones (if there is a biological level, then there must be chemical and physical ones too, but not necessarily vice-versa).

On this view of consilience, sociobiology seeks to explain in evolutionary terms the emergence of such psychological phenomena as altruism, and may well do so on the basis of a gene-centred view, such as kin selection. But this does not in itself preclude the need for concepts and theories peculiar to the psychological levels, arrived at by investigation into the nature of the phenomena peculiar to those levels. It does not commit sociobiology to accepting only gene-centred explanations of psychological phenomena. That would be greedy or eliminative reductionism. Such reductionism can be justified on occasion (we might, for example, show that ghosts are nothing but tricks of the light), but it is not required by the aim for a single system of explanation. A single system of explanation is not the same as a single type of explanation.

As we established in the last chapter, however, what this does mean is that proponents of sociobiology, in common with many other thinkers, will not rest content with the idea that there just are different kinds of explanation for different aspects of phenomena, each with its own use and validity. They will seek to relate them to each other, to look for interconnections, and to try to determine, when there are competing explanations operating at the same level, whether it is possible to eliminate any of them on the basis of their incompatibility with those at some other level which have been fully established. This is not eliminative reductionism, although it does involve the elimination of untenable theories.

Midgley sometimes seems to suggest that there are just disconnected patterns of explanation of this kind, each necessary, and each doing different things (Midgley 1994: pp. 63–70). But she is very far from averse to rejecting some theories, especially in the social sciences, such as Marxism and neo-liberalism, for their inadequacies with respect to their accounts of human psychology (Midgley 1994: pp. 76–7). More positively, her account of the evolutionarily produced forms of human sympathy show how key phenomena of human moral psychology may emerge from the biological levels (Midgley 1994: pp.141–50). It may be that as a matter of fact we will encounter types of phenomena which we can only explain by theories and concepts completely divorced from all the others we have available. But it is not clear that the attempt to produce consilience is in itself a pernicious activity, even if we can have no a priori guarantee of success.

Returning to Hinchman's discussion, we need to note that he attacks the view that Wilson's version of consilience effects a reconciliation between the natural sciences and humanism. It does so, he argues, only by relegating the latter to the task of interpreting and conserving what is already known, whilst it is the task of the sciences to provide new knowledge (Hinchman 2004: p. 22). But, he says, 'interpretation is – or can be – a form of discovery and knowledge'. In explanation of this claim, he says:

> a new configuration of symbolic content (say an ethical theory, a historical explanation, a legal decision, or a fundamental ontology) has just as much inherent claim to be 'knowledge' as does a sociobiological theory that attempts to base ethics on Darwinian adaptation. (Hinchman 2004: p. 23)

His concern is that sociobiology attempts to replace 'critical-emancipatory disciplines' such as philosophy and political theory by biology.

However, this first of all seems to beg the question, by assuming that anything which claims to be knowledge really is knowledge. We are getting close to the 'anything goes' position of Feyerabend, in which all we have are different ways of talking, of which the natural sciences are only one specimen (Feyerabend 1975). At the very least some of the ingredients on Hinchman's list – the ethical theory and the historical explanation, say – need to be compatible with what science has discovered (Noah's Flood won't wash as the explanation of fossils on mountain tops; an ethical theory needs to have some view of 'can' if it is going to suggest 'ought'). Some items on the list – the legal decision, say, or the 'fundamental ontology' – either are not logically the right thing to count as knowledge (a decision is not a knowledge claim) or are simply impossible to assess in terms of knowledge – as are all the claims of revisionary metaphysics, as opposed to descriptive metaphysics (for this distinction see Strawson 1959: pp. 9–11).

Insofar as philosophy and political theory are critically emancipatory it is because they fuse together factual claims, norms/values and prescriptions. On the factual side, it is not clear why biology should not have enlightening things to say – and to enable us to reject conclusively some factual claims, thereby making some norms and prescriptions impossible to accept (showing there are no such things as races makes racism that much less plausible as a normative position; ditto with showing that male and female brains are virtually identical). It is not that sociobiology aims to replace philosophy or political theory, but rather that it aims to make a significant contribution to these disciplines by linking the normative and prescriptive elements to other levels of knowledge.

It is certainly correct to say that philosophy (and the social sciences and humanities) have an unavoidable involvement in normative and prescriptive discourse. Even if biology as applied to non-human life-forms does not have such an involvement, sociobiology certainly does – at least when applied to human beings. As we argued at the end of the last chapter, it is indeed impossible to study how human beings are organised and socially interact without taking a view about how they ought to be organised and interact. To recall the main argument for this, it is because one cannot understand an example of human behaviour unless one grasps the values and norms used by the people whose behaviour it is. One cannot decide whether one has understood these without taking some view about whether or not they can be justified, or made intelligible. Finally, one cannot do that unless one has some view about what actually is a defensible or intelligible value to pursue in the circumstances in question. This line of thought pervades even the most 'scientific' form of social theory, as in the case of Marx's concept of false consciousness.

For these reasons, sociobiology is taking a view of human beings that cannot but have implications for how we should order our lives. But what these implications are is going to be difficult to work out, and subject to all kinds of interactive considerations. It will not be a simple matter of reading off 'ought' from is. If some proponents of sociobiology have suggested otherwise then they have got too simple a picture of their own position. But there is nothing here that stands as a telling objection to naturalism in either ethics or the explanation of human behaviour. The project of sociobiology, via consilience, is a project of linking up levels of knowledge. It is not a matter of abolishing any of those levels. The linking up does not, because it cannot, leave the levels in the same condition as they were prior to the linkage. But it is only by begging all the important questions that one can suppose that the situation prior to the linkage is preferable to what obtains after the linkage.

The objections which Hinchman offers to naturalistic positions such as sociobiology are, however, as much moral as epistemological. He seems to suggest that what is really important in human life is what transcends the sphere of the 'life-process' of society – 'all that concerns physical survival, reproduction, territoriality, labour, etc.' (Hinchman 2004: p. 22). But sociobiology can only deal with the life-process (biological) level – staying alive, healthy and reproducing ourselves. It then reaches up to drag down 'higher cognitive' processes to the 'life-process' level, by giving them 'biological explanations'. This only seems a defensible move, he suggests, because our society has become so corrupt that we have devalued 'experiences that transcend consuming and reproduction'.

This claim expresses a common concern, and a common reaction to any mode of thought which seeks to connect us with the rest of the natural world, rather than to separate us from it as the sole possessors of higher levels of thought and experience, such as art, philosophy, religion and ethics.

The whole picture is, once again, immensely question-begging. It assumes a distinction between the grubby and the refined, the base and the noble, the human and the animal, the material and the spiritual, which is now wide open to objection. One may as cogently argue the opposite, that the 'life-world' is not grubby, base and ignoble. It not something we should seek to distance ourselves from. In particular, the historical story of evolution is simply stupendous, mind-boggling, and utterly fascinating. It may help us to feel at home in the world and to experience ourselves as a unity, not as intellectual ghosts trapped in loathsome physical machines. Again, this is a key idea of Midgley, who has spent much effort in exploring, and seeking to defuse, the fears lurking in the thought that human beings have some deep connections, via their evolutionary history, with other life-forms (Midgley 1979).

Wilson, too, is perfectly willing to accept the importance to human beings of a narrative within which human life can be found to have significance – and a 'sacred' narrative, to boot:

> People need a sacred narrative. They must have a sense of larger purpose, in one form or other, however intellectualized … If the sacred narrative cannot be in the form of a religious cosmology, it will be taken from the material history of the universe and the human species. That trend is in no way debasing. The true evolutionary epic, retold as poetry, is as intrinsically ennobling as any religious epic. (Wilson 1998: p. 295)

Arguably, this work of 'poetry' is one to which Wilson has himself already contributed a great deal. The eminent novelist Ian McEwan is a great admirer of the quality of Wilson's writing. In developing the idea of a literary tradition within scientific writing, parallel to the idea of a literary tradition of the more familiar kind, McEwan finds a place for Wilson's powerful and lyrical interweaving of scientific theorising about, and descriptions of, the natural world (McEwan 2006). He particularly cites Wilson's descriptions of the Amazon rainforest and of the world of the soil from Wilson's *The Diversity of Life* (1992). In the face of these features of Wilson's work it is harder to sustain the view that sociobiological approaches to the understanding of human life necessarily plunge us into an arid reductionism.

One key concession does, however, have to be made to Hinchman with respect to this exposition of the role of narrative in human life from the sociobiological perspective, namely that it is a clear implication of that perspective that there is one true narrative to be told, and that this narrative is destined to oust all others from the repertoire. Hinchman, as we noted, believes that the ability to formulate narratives within which different human individuals and groups can find their own distinctive forms of meaning and purpose is crucial to human freedom. It has been argued earlier that fixity of theoretical perspective does not in itself necessarily limit in any objectionable way the possibilities for human beings to develop new modes of understanding. However, it is an important issue whether universal acceptance of the evolutionary narrative would be found to exercise a limiting or deadening effect upon the possibilities for human beings to develop new forms of understanding, whether of themselves or others. Would something priceless be lost if the evolutionary narrative became the dominant form of human self-understanding?

It has to be admitted, also, that this narrative has its extremely dark side too – it is a tale of life-forms emerging by an immensely long process of death, suffering and destruction. Certainly, as Lisa Sideris has argued (Sideris 2003) the evolutionary approach to our self-understanding has some extraordinarily challenging claims for us to contend with – that the process has no pre-ordained end-point or purpose; that nature does not contain harmony and balance; that death and suffering are integral to it, not optional extras; that human beings can rely only upon each other for care, concern and compassion, and that these may have a fragile basis in our natures. The most troubling point of all is that our deepest moral concerns emerge solely and contingently from our evolutionarily produced natures, not from some fundamental source within the very fabric of being. These are not just deeply disturbing thoughts, they are also very new ones, and will take a great deal of time to assimilate them and rethink our philosophical positions in the light of them. We will have to look more closely at these implications in subsequent chapters.

But there is no reason to suppose that a correct or defensible response to them is to seek to re-establish an untenable form of dualism which enables us to salvage what we most value by divorcing it from our evolutionary history. If the evolutionary story is correct (and Hinchman seems to accept that that is the way it is looking at the moment) then we have to accept it and integrate it into our philosophy, art, ethics and spirituality. Arguably it is only a form of dualism which was developed during the long centuries when we accepted the 'skyhook' accounts of this world and our place in it that leads us to look down on the 'life-world'.

Now we have good reason to adopt a completely different picture we can no longer take those older views for granted. Humanism, if it is committed to this kind of dualism, was a child of its times. Its fundamental values remain of crucial importance. We now have to find a way of supporting those values in the light of the new view we have been developing of ourselves. Hinchman has not given us yet any reason to suppose that that is impossible. It is not a claim of evolutionary theory or sociobiology that we 'cannot in any sense transcend our hominid ancestry'. He has erected a straw man.

However, Hinchman goes on to make what appears to be a further telling objection to any attempt to link up an account of human ethics, and environmental ethics in particular, to the evolutionary perspective. He argues that it is very hard to see how environmental ethics can be deduced or explained on the basis of natural selection. After all, environmental ethics argues for the inherent worth of nature, and prescribes biodiversity or wilderness preservation even when our own survival or economic interests may be jeopardised. He notes that arguments for biodiversity offered by Wilson appeal to the importance of the natural world for the spiritual, aesthetic and moral dimensions of human life, not just to the narrowly self-interested causes of survival.

But this appears to ignore how evolutionary theory has come to account for the basis of morality and altruism in human beings (see, for example, Ruse 1986; Ridley 1996). The problem which the latter phenomena pose for evolutionary theory is that of explaining how any human individual (or any other individual organism) can take account of the interests or worth of any other. As we have frequently noted, in the human case, kin selection and reciprocal altruism can play a vital role in the explanation. Ridley has also cited the phenomenon of sexual selection as a possible source of love for offspring and spouses (Ridley 1996: pp.133–5). These are said to provide the biological basis for morality. They explain how human beings can be genuinely altruistic whilst still being purely evolved beings. They also explain why that altruism will be harder to extend to non-kin, non-offspring, non-group members. In such cases the intellectual virtues, such as consistency, may come in to remedy the gaps. But they will be less strong as sources of motivation – and they observably are (Baxter 1999: pp. 40–41).

Environmental ethics is a very complex activity, appealing, as do all ethical arguments, partly to emotional and partly to intellectual considerations. But there is no a priori reason why evolution by natural selection should not have provided within the human brain tendencies and dispositions that can form the basis for ethical systems involving the attribution of value and prescribing obligations, towards the non-human. Whether there is in fact any reason drawn from evolutionary theory for believing otherwise is a matter to be investigated in a Chapter 8, when we turn to consider Wilson's version of environmental ethics in greater detail.

Clearly, human ethics, including environmental ethics, are data to be explained by evolutionary thought. They cannot be ignored. This means that if evolutionary theory cannot explain ethics it should be rejected as, at least in important respects, inadequate. But the theory does try to explain these phenomena. The explanations may not work, but then evolutionary theorists will have to try again. The theory should be given time to make its attempts. Only if evolutionary theorists were refusing to accept that there was anything in human ethics to be explained could environmental ethicists properly reject it.

However, many moral thinkers believe that there are serious problems in any attempt to explain morality in naturalistic terms, and that Darwinian naturalism of the kind propounded by the sociobiological approach is no exception to this. We need now to consider this critique of naturalism in ethics more directly, which is the task for the next chapter.

Chapter 7

Naturalism and Morality

The last chapter ended with the thought that Darwinism needs to account for morality, and that if it could not do so, then it would need to be rejected as inadequate, at least as an explanation of the human case. We also noted that the explanation of morality that Darwinism does supply is a troubling one, at least for those – the majority – whose view of morality has been formed on the basis of traditional religious and humanist views of the matter.

The trouble stems from the naturalism of the Darwinian perspective on morality. This perspective encompasses a variety of considerations, including the thought that, as it was expressed in the last chapter, our deepest moral concerns emerge solely and contingently from our evolutionarily produced natures, not from some fundamental source within the very fabric of being – a transcendental source which lies beyond the contingent, causally connected world of experience. Of course, it is not only morality which is often held – by religious believers, humanists and others – to have such a source. Religion itself is also oriented towards the supernatural, which is another name for the transcendental. We will have to consider more directly in Chapter 9 the implications for religion of the Darwinian worldview.

However, in this chapter we will consider more fully the naturalism of Darwinism in connection with morality. This will then lead on in the next chapter to an examination of how far Darwinism can accommodate a robust environmental ethic. Such an examination will require attention to be paid to the possibility that the account of morality that Darwinism produces makes it very problematic to suppose that human beings are able to grant any moral importance to the non-human world. If this possibility can be shown to be actual, than it implies that there is no real connection between the two sides of the Darwinian worldview identified at the start of this book. That is, it will not be true that viewing human beings as part of nature transforms both our understanding of that human nature and underpins a moral view that takes us to have moral obligations towards the non-human.

Wilson has provided an examination of the case for moral naturalism – what he refers to as 'empiricism' – and against its transcendental competitor, which leads us to a consideration of the central issues. In Chapter 11 of the book which we have already looked at in connection with consilience and its connection with the social sciences, *Consilience: The Unity of Knowledge*, Wilson considers how sociobiology accounts for morality and religious belief. The first issue that needs to be addressed in this area concerns how we should conceive of moral principles and precepts. As Wilson expresses the issue in his opening sentence,

> Centuries of debate on the origin of ethics come down to this: Either ethical precepts, such as justice and human rights, are independent of human experience or else they are human inventions. (Wilson 1998: p. 265)

He immediately, and correctly, notes that the divide between supporters of the former view – transcendentalists – and those of the latter – empiricists – does not coincide with that between religious and non-religious points of view. It is possible to believe that moral precepts are independent of the human mind – discovered by humans rather than invented by them – and differ over whether they originate in God's will, or are also independent of that will (Richards 2000: pp. 188–92). In fact, there is a subsidiary debate here which Wilson ignores, between those religious believers who think that God is the origin of morals and those who believe, with non-religious transcendentalists, that even God has to recognise the good, rather than invent it. The latter position arises within religious views because the former position seems to have the unfortunate consequence that contingency with respect to morals, from which transcendentalism is supposed to offer an escape, is simply raised again at a higher level. If God invents moral precepts, rather than recognises them, then it seems that He could have simply inverted the current ones, making what is currently viewed as evil into good, and vice-versa. This seems to be inconceivable. Morality does seem to have a given, substantive, content. Transcendentalism thus seems to be driven into a Platonic position, in which there is, in some way, an objectively existing realm of moral 'forms' which minds – finite or divine – can do no more than acknowledge.

Wilson in effect recognises that the case for and against religious belief is really not the main issue here. Both secular and religious transcendentalists postulate an objective moral order: 'In short, transcendentalism is fundamentally the same whether God is invoked or not' (Wilson 1998: p. 266). However, arguably he does not really get the issue between empiricists and transcendentalists quite into accurate alignment. The point of dispute between transcendentalists and empiricists is not that the former believe in objectivity of moral precepts and the latter do not. Rather it is the contrast between necessity and contingency which is the real heart of the dispute.

We can see this when we consider Wilson's mode of introduction of the empiricists' position:

> Ethics, in the empiricists' view, is conduct favored consistently enough throughout a society to be expressed as a code of principles. It is driven by hereditary predispositions to mental development … causing broad convergence across cultures, while reaching precise form in each culture according to historical circumstance. (Wilson 1998: p. 267).

This is the gene-culture co-evolution account, of course. It specifies that we need to know which epigenetic rules are operating so as to predispose human beings towards the acceptance of certain culturgens that form part of our repertoire of moral behaviour, and the effect which particular forms of social environment play on the actualisation of such epigenetic rules in given instances. To understand the prior genetic production of the epigenetic rules we need the standard account of natural selection and inclusive fitness. The epigenetic rules must be adaptations, allowing their possessor an advantage in the handing on of the genes to descendants.

The point about this argument, whatever its merits or demerits as an account of the origins of morality, is that it is as perfectly objective an account as that of the transcendentalists. It has to rest on knowledge, not mere opinion, as Wilson emphasises: 'The importance of the empiricist view is its emphasis on objective knowledge' (Wilson 1998: p. 267). Where it differs from that of the transcendentalists is that it this knowledge appears to be knowledge of contingency. That is, we have to be able to discover the particular circumstances within which the human species developed which give an evolutionary point to certain predispositions to behaviour. Since evolutionary accounts are all contingent, this implies that if the environmental circumstances of the human species had been different, certain other predispositions would have become embedded as epigenetic rules, and a different set of moral precepts would have emerged as the universal underpinning of various cultural variations.

It is this underlying contingency that causes the 'nausea' (to use Sartre's term) which may be felt by transcendentalists of all persuasions at this point. They hanker after a sense of necessity – that in some way the basic content of moral precepts, however vaguely they might be expressed, must have a certain content, on pain of ceasing to be morality at all. Of course, this subjective sense of the necessity of morality can be explained, as Wilson does explain it, as the awareness of the 'moral sentiments' favoured by eighteenth century moral philosophers such as Hume, Hutcheson and Smith (Wilson 1998: p. 280). On this account the subjective sense of necessity is an awareness of the operation of the epigenetic rules which are genetically produced within our brains.

However, it may be possible to provide a justification, rather than simply an explanation, of such a sense of necessity in a way that is compatible with the contingency-based approach of empiricism. It may be possible to formulate an argument based on the idea of the 'forced move' which we have already noticed as a possible source of fixity in human life in an earlier chapter (see Chapter 2 above). Thus, it might be the case that, for any intelligent species with the social characteristics of human beings, certain forms of co-operation are essential to the survival of that species in the conditions approximating to those of human beings on this planet. As Darwin puts the point:

> The following proposition seems to me in a high degree probable – namely that any animal whatever, endowed with well-marked social instincts ... would inevitably acquire

a moral sense or conscience, as soon as its intellectual powers had become as well, or
nearly as well developed, as in man. (Darwin 1901: pp. 149–50)

Wilson provides us with an example of how such a 'forced move' argument
might be deployed, in a hypothetical example of a Palaeolithic hunter band. In
this example, each individual hunter ponders the relative advantages of hunting
on his own (he is less likely to be successful than if he cooperates with others; all
the meat will be his if does succeed, but he will cause animosity by lessening the
chances of success of the others) and hunting in a group (he is more likely to be
successful; he contributes to the success of others and thus gains approval, but
he will have to share the kill). He estimates the latter to be the preferable strategy,
and what is true of him is true of the others. In another example of the Baldwin
effect, if there is a genetic tendency to cooperate it will tend to become widespread
in such circumstances (Wilson 1998: p. 283).

Generalising from such examples, we may conclude that no alternatives,
objectively speaking, give members of that species, in those conditions, a better
way of ensuring that they pass on their genes to their offspring. This will be a
forced move, objectively thrown up by the contingent combination of species and
habitat. The sense of necessity attaching to the predispositions embodied in the
moral sentiments will thus be justified, even though it is an entirely contingent
matter that organisms that benefit from just those predispositions come into
existence.

This may not be quite the sense of necessity which transcendentalists hanker
for. It is a kind of relative necessity – necessity relative to initial presuppositions.
What transcendentalists seek is absolute or categorical, necessity. Goodness
is goodness, full stop. Here, it seems, is a genuine parting of the ways between
transcendentalists and empiricists. Empiricists (or what I think we should call
naturalists) can only get as far as relative necessity, and that is not what is
required by transcendentalists. The problem for transcendentalism is, however,
that it is very difficult to account for absolute necessity whenever it is appealed
to. This, after all, is the problem that has bedevilled the ontological argument in
its various forms – the idea that the existence of God can be shown in some way
to be absolutely necessary – that God could not fail to exist. The problem is that
the non-existence of God does seem to be a possibility that can coherently be
entertained, otherwise atheism ought to be as clearly inconceivable as the idea
of a three-sided square.

The naturalist can, however, account for the belief in the transcendental origin
of moral precepts, and thus in their absolute necessity. To summarise Wilson's
version of this account, such a belief is said to come about as the result of a
transformation of what originate as epigenetic rules firstly into socially-formulated
and enforced versions of moral predispositions – that is, into laws – and then,
usually with the aid of religious conceptions, into supernaturally sanctioned
absolutes (Wilson 1998: pp. 278–9). Human beings, in other words, are the origins
of these conceptions, not some transcendent or supernatural source. By contrast,

the transcendentalist is, as we have noted, faced with a more difficult, and perhaps insoluble, problem, of accounting for the existence of such absolutes without making any appeal to the contingent and the empirical. As we have seen, what this results in is sheer assertion, rather than any explanation.

The naturalist account does not have to postulate the idea of absolute necessity. It can provide a version of necessity which at least may go some way towards justifying the common sense of how moral claims function, and can explain the subjective feelings of necessity on the basis of the awareness we each possess, perhaps to different degrees, of the epigenetic rules with which our genes equip us. What it does not do, and given the logic of its position, cannot do, is provide a vantage point outside of the evolutionary perspective from which it can judge the outcome of evolutionary processes in a moral sense.

Thus, it is possible that the 'forced move' argument mentioned above might no longer be applicable because of some radical change in the human environment. The future evolutionary development of mankind may take it in a direction such that inclusive fitness is best achieved by the alteration of the epigenetic rules currently underpinning what we regard as moral behaviour. Humanity may evolve away from morality, hard though that is to conceive at the moment. The only point of view from which that will be lamentable, given the logic of gene-culture co-evolution, is that which human beings, by and large, currently have. We will be bound to feel that in such a future, human beings will have lost what is central to their humanity. Given our current natures such a prospect will necessarily appear nightmarish.

From the point of view of the future human beings themselves, experiencing their natures from the inside, a sense of necessity and fittingness within their tendencies to act in what we think of as immoral or non-moral ways will be, presumably, as compelling as the moral sense currently is for most human beings. Gene-culture co-evolutionary accounts of morality provide no basis for gainsaying this judgement. However, it is also part of the logic of that position that there is no basis for criticising our current, moral, dispositions. We are what we are, and cannot simply choose to become such (from our point of view) moral monsters. Deep-seated changes will be needed in our environment and epigenetic rules to make such a form of life appear both appropriate and sustainable.

Hence, the alarm we are inclined to feel at such a hypothetical prospect is primarily an intellectual one – the discomfort already noted which arises from a theory-based recognition of contingency at odds with our experience of a sense of necessity. It cannot plausibly be based on the claim that gene-culture co-evolution theory threatens to undermine the bases of human morality. For there is no warrant to hold that this theory supports the conclusion that human moral impulses are fragile just because they have been arrived contingently by the fortuitous route which evolution has in fact followed. Rather, the reverse is the case. If the human condition retains most of its current features and our human natures are as anchored in epigenetic rules as gene-culture co-evolution theory suggests, then the prospects of humanity evolving into, say, purely Hobbesian individuals are remote.

The reference which we have been making so far to the unavoidability of certain forms of experience in the moral sphere also plays a crucial role in another long-standing dispute between naturalists and transcendentalists, raised in the last chapter in the discussion of humanism. This concerns the transcendentalists' claim that morality presupposes the existence of moral actors, and that actors of this kind, such as human beings, possess a fundamental kind of freedom. This is the freedom to choose between courses of action, to choose the good or the bad, to choose to do what ought to be done, or to choose not to. A free agent of this kind is capable of self-rule, self-control and self-criticism. Such an agent can reflect on its choices and criticise its emotions and feelings, rather than simply act upon them. This is the autonomous self, celebrated by Kant and the Judeo-Christian–Islamic tradition, as well as being central to humanism, as we saw in the last chapter (Kant 1958: p. 114).

It is important for this conception of freedom that what human beings do should be conceptualised as freely chosen action, not simply as behaviour. Behaviour is a matter of simple doing. Living organisms all behave in various ways, and this behaviour can, in principle, be explained purely in causal terms, as responses to various internal and external stimuli. Moral agents act for reasons, which they have formulated in some sense, and which may be used to justify the action, or condemn it, for other moral agents' benefit, or in the course of the agents' own later ruminations on the actions which they have performed.

This conception of moral autonomy has long been used to differentiate human beings as a different kind of being from all other animals, even if it is conceded that they have shared a common descent with them. It has been held to imply, as Kant emphasised, that the causal explanations of behaviour, of the kind which are essential to the sociobiological approach to the study of all other animals, are simply inappropriate to the understanding of human morality (Kant 1958: p. 56). Kant, of course, proposed a 'two standpoints' solution to the understanding of human beings as moral actors. We are, qua material beings, embedded in causal networks along with all other material things and can view ourselves in this light. But qua moral agents we must view ourselves in a completely different light – as separate from all such causal networks and possessing the self-determining status of rational actors (Kant 1993: pp. 113–14).

This, then, is the second facet of moral phenomena – moral agency – which seems to many to point inexorably to the transcendentalist position, rather than to the naturalist/empiricist one. Putting the two points together, we can say that on the transcendentalist view, both the agents of moral action and the principles upon which those agents act, have to be granted transcendental status. Religious believers give this status a specific ontological interpretation – it requires the postulation of a supernatural realm, a realm in which purely spiritual beings may be located. Non-religious humanists do not go in this direction, remaining content simply to maintain that no naturalist explanation of the kind favoured by sociobiology can be appropriately deployed to account for moral phenomena, properly understood. The freedom and dignity of human beings requires that naturalism in morality be rejected, and with it any hope of a workable sociobiological explanation of human beings, given the centrality of moral concerns to human life.

It is pertinent to note at his juncture that the transcendentalists' account of morality involves an interesting interplay between the ideas of freedom and necessity. On the agency side, moral agents are free to choose how to act. They can choose to do the good, or not. But if they choose to act morally, then they are faced with certain inescapable requirements, for on this view the right and the good are specified by certain absolutely necessary principles.

One possible complication in the transcendentalist position should also be noted, namely that the arguments for the transcendental status of agency and principles are logically independent of each other. That is, it is possible to maintain the transcendental standpoint with respect to either of the two positions while rejecting it for the other. Thus, we could argue that moral principles are necessary, but that human beings do not possess the freedom of moral agents, and are induced to follow the principles by purely causal processes. Or we could argue that human beings are autonomous beings whose actions cannot be explained satisfactorily in purely causal terms, but that moral principles are all contingent human inventions, having at best the relative necessity deriving from the 'forced move' considerations mentioned above.

Religious believers will be inclined to support both transcendentalist positions, while it is not difficult to find humanists who argue for human freedom and autonomy but are willing to support the idea that moral principles are in some way a human invention. It is not so easy to find supporters of the view that human beings can be analysed purely in causal terms, while moral principles are necessary. This suggests that it is the claim of human autonomy and dignity that is in effect the dominant element in the humanist transcendentalist position.

By contrast, it looks as though the naturalist tradition has to dispense with absolutely necessary principles, relying instead only on the possibility of relatively necessary ones (given the forced move argument). It also appears that its evolutionary account of the origins of morality does commit it to the conceptualisation of moral actions as simply a form of behaviour subject to causal explanation in terms of the interaction of epigenetic rules and culturgens. Does not this mean the loss of the idea of the free moral agent, which seems to be such an indispensable part of our understanding of morality, and with it the priceless idea of the dignity of human beings? Does not sociobiology reduce human beings to just another species of animal – albeit a complex one? May it not fairly be accused at this point of greedy reductionism – of showing that our sense of ourselves as dignified, free, moral agents, responsible for our lives and our choices is an illusion? Are our actions 'nothing but' the results of complex causal operations within our brains, analysable by the equations of gene-culture co-evolution?

One way to approach these issues is to consider a somewhat different version of the 'two standpoints' answer formulated by Kant to account for our sense of ourselves as moral agents. Instead of formulating the two standpoints as 'actor' and 'patient' we should focus upon the two standpoints represented by the 'internal' and 'external' perspectives with respect to human action. All human beings that are mature enough and sane inevitably have the internal perspective with respect to their own actions. We all possess awareness of our actions as our actions, issuing from our own decisions which have often involved the weighing up of pros and cons, consideration of the interests affected and so forth. We cannot avoid this internal perspective. In our own case it is the only perspective available to us directly, that is, unmediated by some other, prior, form of awareness.

The external perspective, by contrast, is just as inevitably the one we automatically adopt with respect to other actors. We observe their behaviour as something to be interpreted, understood and perhaps misunderstood. This perspective is the foundation upon which we erect both folk psychology and the more theoretically loaded versions of explanatory schemata such as those of human sociobiology.

With respect to ourselves, the internal standpoint is adopted automatically. We have to work – sometimes very hard – to attain the external standpoint as applied to ourselves. This may simply involve the attempt to see ourselves as we appear to others: 'O wad some Pow'r the giftie gie us *To see oursels as others see us!*' in the memorable lines of Burns's *To a Louse*. It may involve trying to see our actions and behaviour as embodying some pattern or exemplar into which the behaviour of others may also be fitted. Ultimately such a perspective may lead on the theorisation of human behaviour in abstract, general terms, in which the internal standpoint has gone completely.

With respect to other agents, the difficulty is the opposite one. It is automatic that we have the external standpoint with respect to them, and then we slip easily into the categorisation of their observed behaviour, insofar as we understand it at all, in terms of general patterns in which a causal network is easily applied. This does not have to be a matter of restricting ourselves solely to what can be seen. It is part of the same general external approach that we interpret the behaviour in terms of mental phenomena, such as beliefs, emotions, values and so on. Indeed, the descriptors we use for the categorisation, and understanding, of the behaviour of others embody the mental aspect of the whole phenomenon. We say someone is blushing, for example, only if we take the observable phenomenon of skin reddening as connected, via what Wittgenstein referred to as the criterial relationship, with a feeling of shame or embarrassment (Hacker 1972: pp. 283–309).

To adopt the internal standpoint with respect to another person is a rather more taxing activity than this. It involves an attempt to re-enact the thoughts and feelings as experienced by the actor within one's own consciousness. This takes effort, and probably a certain amount of aptitude that may not be evenly distributed amongst all human beings to equal degrees, or to equal degrees at different stages of their lives. We sometimes have to work to attain the internal perspective with respect to the actions of others. Literature and other forms of dramatic representation can help enormously in this regard.

What is clear is that the sense of ourselves as free and autonomous beings, as acting for reasons we have chosen to attend to and have weighed up by our own mental efforts, is primarily attached to the internal perspective. This is why the appeal of the transcendentalists' position on this matter is so powerful. When we consider the alternative appeal of the empiricists' position it is natural to fall automatically into the external perspective, and to think of the actions of others. That is the perspective from which the categorisation of behaviour, the finding of general patterns, and thus the postulation of causal explanations, including those couched in the terms of evolutionary theory, is found to flow most easily. It is only when we are challenged with the question of whether we are happy to think of our own behaviour in this way that a dissonance is immediately felt. The idea of our actions as being caused by general processes of which we are an exemplar clashes directly with our experience of ourselves as autonomous beings.

Does this mean that one or other perspective has to be given up as an illusion? There is no reason to speak of illusions in this case. The experience of ourselves from the internal perspective is not an illusion – not if 'illusion' means something which does not really exist. Our sense of ourselves as deliberating, feeling emotions, being torn in different directions, making up our minds and so forth really does take place. But it is also true that much of what we do can be understood as the kind of thing which certain kinds of human beings tend to do in similar circumstances. We are never unique, at least to the degree that we often take ourselves to be. That is why the external perspective is available at all. It is what enables us to understand others, and thus to understand ourselves. It is often a considerable relief to find out from others that they had the same feelings that we had, that they were moved by the same considerations, that we behaved and acted much as others would have done in such contexts. The relief does not require that everyone would do the same in comparable circumstances. All that is needed is the sense that we are part of the human mainstream, the normal range of variability – not freakish. We may wish to be regarded as autonomous human beings, but the emphasis is upon the 'human beings' as much as on the 'autonomous'.

What all this implies for the dispute between transcendentalists and naturalists over moral agency is that naturalists do not have to diagnose the transcendentalists' emphasis on moral autonomy as based on an illusion, and thus as to be rejected. Rather, they need to challenge the understanding of this phenomenon. They should regard it as an inevitable part of the unavoidable internal viewpoint – something we could not abandon even if we tried. It captures an essential part of the experiential landscape of self-conscious, self-aware beings such as ourselves. But it is not inherently incompatible with the provision of a causal explanation, of the kind sociobiology proffers, of the phenomena of decision-making, deliberation, agonising over alternatives, and the rest. Rather, it is only because we have available such causal explanations that we can fully understand others, and thus understand ourselves – which is part of the lesson of Wittgenstein on the problem of other minds and private languages.

For we quite happily, though not, of course, infallibly, categorise the actions of others in general terms and under general headings, in ways which imply causal connections, in the course of using our 'folk psychology' on each other. This too is indispensable to the conduct of our lives. Hence, it cannot be egregious for sociobiology, or any other general theoretical perspective, to do the very same thing by means of categorisations derived from areas beyond folk psychology, such as evolutionary theory. Those theories may be wrong, but they cannot properly be convicted of outraging human dignity by the very nature of what they are attempting.

We have noted, in any case, that even on the account of most transcendentalists the content of moral principles is necessary. Morality is held to enjoin certain general aims and values by the very nature of what it is supposed to be. Hence, the autonomy of human moral agents, on this account, is not supposed to extend to the matter of freely deciding what is to count as a moral requirement or consideration. It will, however, extend to the matter of weighing up competing moral considerations, and of reaching a conclusion which will probably be different from that reached by others, but the same as that of others still.

What this implies is that the autonomy of the individual moral agent is not quite as giddily transcendent as it might appear at first sight. After all, even on Kant's version, there is the expectation that once rational beings have applied the categorical imperative to the proposed courses of action between which they are choosing they will reach the same conclusion about what they ought to do.

The empiricists understand the sense of autonomy of the agent as deriving from the experiential nature of the internal viewpoint, but regard that as explicable in causal terms from the external viewpoint. They will hold the content of the moral precepts as possessing only relative, not absolute necessity. Their naturalistic account of the emergence of moral norms, and the epigenetic rules that embody them, will explain how the human species gets to contain moral agents in the first place. It will not, however, lead to the transformation of ethical reasoning into a predictive science, for the epigenetic rules, as we saw in the course of the discussion of gene-culture co-evolution theory, do not have to be narrowly constraining. The role of cultural variants and the looseness of fit between epigenetic rules allows for the individual to weigh up and decide complex moral matters in a way which reflects their own particular evaluative stance – tough or tender-minded, for example – and cultural milieu.

There are no absolute criteria against which moral values and principles will have to be tested and perhaps eliminated, on this view. The meta-ethical theory which best fits with the empiricist view is that of immanentism, as developed by Michael Walzer, for example (Walzer 1983: p. xiv). This implies that the development of moral thought, and the phenomena of moral critique and transformation, are best understood as proceeding from already established value positions. They will involve the intellectual exploration of the implications of those commitments already undertaken, with an eye to extending them in new ways, to teasing out hitherto unnoticed implications and to the use of some of them to reject and/or amend others. Walzer accepts that there are some fixed positions in the moral universe, but holds that they are too general and abstract to be of much help in the conduct of moral debate. The empiricist/naturalist standpoint could accept this happily if empirical investigation did not succeed in unearthing any extensive and complex epigenetic rules. But it might be that there will turn out to be more, and more substantial, fixed points in the moral sphere than Walzer allows for.

Another moral concept that finds a natural place within the naturalist view of morality is that of integrity. In many ways this holds a place within the empirical view of morality akin to that which autonomy holds within the transcendentalist viewpoint. In the course of a passage in which he is characterising the empiricist viewpoint Wilson puts the following claim into the mouth of an empiricist:

> True character arises from a deeper well than religion. It is the internalization of the moral principles of a society, augmented by those tenets personally chosen by the individual, strong enough to endure through trials of solitude and adversity. The principles are fitted together into what we call integrity, literally the integrated self, wherein decisions feel good and true. (Wilson 1998: p. 274)

We need to remember, of course, the co-evolution account of where the 'moral principles of a society' derive from. What this suggests is that integrity is a matter of being true to what one is, as at bottom an evolved moral being. Of course, to repeat the point made at the start of this chapter, from the naturalistic perspective this status of the individual has no absolute, transcendental importance. It is basically a matter of contingency that human beings have evolved to have this moral character. From the external viewpoint it is just a fact about human behaviour that human beings behave in certain specific ways, dubbed moral, in certain circumstances. This fact can be explained in terms of co-evolutionary theory.

But from the internal viewpoint, being true to what one is, making decisions based upon what are experienced as one's profoundest feelings, or best articulated value positions, is the heart of being a person of integrity. People lacking integrity, on this view, have no strongly developed sense of what they are, and are easily swayed by the strongest psychological forces in play within their minds at any given time. Perhaps few people are ever in this condition, or in this condition for long periods of time. From the external viewpoint they are people of whom one says that one never knows where one is with them.

On this account of integrity, of course, one could coherently regard someone as possessing integrity in this sense even though they subscribe to principles one finds abhorrent. But then one may still feel a grudging admiration for the fact that they say what they mean and mean what they say, even if what they say strikes one as being highly objectionable. This suggests that integrity, although important for the individual self, and for others, is not the only thing that counts. But then the same can be said for freedom and autonomy. It is possible freely and autonomously to choose to do wrong, to pursue morally abhorrent ends.

The upshot of this discussion is that the empiricist/naturalist position in morality can be shown to be internally coherent, to explain, without explaining away, important phenomena, such as the sense of autonomy and human dignity and to account for at least some of the elements of necessity which are involved in moral thought. We have seen the ways in which the case for transcendentalism in ethics, with respect to moral principles and moral agency, can be shown to be avoidable, and inherently problematic. Naturalism in ethics is thus not committed to a debunking, greedy, eliminative reductionism in its discussion and explanation of morality.

However, it has long been recognised that the sociobiological forms of naturalism imply that human moral psychology bears the mark of its evolutionary past. In the case of inter-human moral thought and action, this means that there may well be general limitations on our capacity for human sympathy deriving from the origins of human moral thought and feeling in the small-group ambience of our forbears. Further, there is the possibility of natural variation between individuals in the degree to which they are capable of feeling sympathy with others. It is, of course, on this view an empirical matter to discern how far this is so.

Natural variation among human beings in the degree to which their moral psychology is constrained in this way may well lead those with larger natural sympathies to try employ techniques of immanent critique to seek to move their more initially constrained fellows in a more extended direction. But, if the limitations are based on epigenetic rules, then they may have difficulty in doing this successfully. This in turn points to a deeper worry many will have with the gene-culture co-evolutionists' position, namely that such supposed variability in the capacity for moral sympathy could justify the claim that certain individuals are just 'better at' morality than others. Just as musical aptitude varies and we do not reasonably expect everyone to have it to a high degree, indeed some people suffer from tone deafness, so, if moral sympathies derive from epigenetic rules which show variability, is it reasonable to regard everyone as subject to the same moral requirements? Indeed, does it make sense any longer to criticise someone for limited moral sympathies, if the limitation has an epigenetic basis?

Well, such variation does seem to be a fact of human nature, however we account for it. There are, at one end, psychopaths who appear to be devoid of much human sympathy, and at the other people capable of feeling intense sympathy with others all the time. Both types are problematic for others. In between these extremes what we in practice reasonably expect is a basic minimum of fellow feeling between human beings which may be said to underpin basic decency with respect to each other. Beyond this, the requirements of morality are a subject of endless dispute, partly due to the variations in sympathy already alluded to and partly due to more obviously cultural factors, such as the specific versions of moral norms to which different societies have given their support.

These everyday phenomena of the moral life are the subject of debate between sociobiologists and culturalists insofar as the former are more inclined to posit a genetic basis for some of the variability in human moral sympathy, whereas the latter are more inclined towards explanations in terms of different modes of socialisation. However, this version of the nature/nurture dispute leaves the questions we raised above unanswered. However the variability mentioned is accounted for, it does exist, and it does seem to have a basis in the settled character of particular human beings, however that character has been formed.

In the case of morality we do not employ the language of talent and aptitude, as we do in the case of aesthetics. We must hold each other to certain standards in the moral case because of the role morality plays in human life. The aim of morality is, at least, to govern conduct between all the members of a society, whatever the limitations of their sympathies or other characteristics important for the conduct of interpersonal relations.

This is not an unusual phenomenon. Another example comes from military life, where all warriors are required to adhere to certain minimum standards, such as following orders to kill the enemy even at the risk to their own lives, even though it is well recognised that the military virtues of courage, aggression, cunning and resourcefulness are not equally distributed amongst all the participants. Those responsible for military affairs, of course, often seek to form elite bodies of fighters who do possess military virtues to a high degree, but such bodies in the nature of things can only form a relatively small part of the total. Such a possibility derives from the fact that military activities have a specific function, and it becomes possible to select individuals who are well equipped to carry out that function. In the case of morality, there is no such specific function, so that the idea of recruiting bodies of those with a particular talent for morality has no obvious sense.

What this means is that the fact that it is an implication of gene-culture evolution, and other sociobiological forms of naturalism, that there are likely to be variations within the human population with respect to morally-important traits and characteristics is not the serious difficulty which it is sometimes thought to be. Once we remember the point of morality within human life we can see that such variability does not licence a similar variability in moral requirements applied to mature, sane individuals. Some of these may find it easier than others to meet the requirements of morality, for various reasons, of which the particular contour of their epigenetic rules may be one. But we have to hold all to at least some minimum standards if we are to have morality at all.

It is important to recognise that although defenders of naturalism are not bereft of replies to their transcendentalist critics, and have important criticisms of their own to offer of their opponents' position, this is an issue which cannot be said to have been conclusively resolved yet. As Wilson argues, although the naturalists have gradually been gaining ground against their transcendentalist opponents, they cannot yet claim complete vindication of their position. Further, the issue of morality is crucial for the whole project of consilience. If naturalism in general, and its Darwinian version in particular, cannot successfully link up morality to biology, then the social sciences and humanities will forever remain disconnected from the rest of human knowledge (Wilson 1998: p. 288). It is worth bearing this point of Wilson's in mind, for he is someone who is often accused of a kind of dogmatism with respect to sociobiology which his actual writings frequently belie.

All of the points made in this chapter with respect to the criticisms made by naturalists and transcendentalists of each others' positions have focused exclusively upon the issue of morality as it concerns the actions of human beings with respect to each other. But it will be recollected that one of the main aims of this book is to consider how far a connection can be made between the Darwinian form of naturalism and the position of those environmental ethicists who take the view that the evolutionary account of human beings, involving their common descent with other life-forms and their ecological interconnections with them, have important implications for the development and content of environmental ethics.

It is a serious issue as to whether sociobiology, especially in its co-evolutionist form, gives us any good reason to suppose that human beings can, in a whole-hearted and concerted manner, extend their moral concerns beyond the human species entirely. This is the question to which we will turn in the next chapter.

Chapter 8

The Possibility of Environmental Ethics

Having addressed in the last chapter the general issues surrounding the possibility of naturalism in ethics, let us now return to the issue introduced, but not explored, in Chapter 6, which is the connection, if any, between the Darwinian worldview and environmental ethics. Let us recall the key claim of much contemporary environmental ethics, that, since human beings are a part of the natural world, not set over against it, our moral concerns should encompass the interests of other – perhaps of all other – inhabitants of that world. The problem with this as it stands is that former claim – we are part of the natural world – does not in itself logically imply the latter one – we have moral responsibilities towards that natural world. As we have seen, the former claim can be put to use in the argument that human moral thought and action can be given an evolutionary explanation, and thus a naturalistic basis. But the attempt to show that human morality arises from our evolutionary past is in itself insufficient to establish that we have any moral responsibilities towards the non-human beings with which we share an evolutionary history.

Indeed, it seems plausible to argue that that very argument shows that we should expect human morality to be directed more readily in certain directions than in others – towards human kin and loved ones, rather than towards human strangers or the non-human. Our evolutionary history may mean that the rest of the natural world may not come very readily to the human mind as an object of possible moral concern. Hence we appear to need some further argument to show both that we should take a moral account of the non-human, and that we have the kind of nature which will allow us to do so to some appreciable degree.

Wilson is one of the few sociobiologists to have devoted extensive thought to the defence of the claim that our membership of the natural world commits us to an extensive set of moral responsibilities with respect to that world. As we will be discovering, the two areas of sociobiology and environmental ethics are interconnected in his thought. In his view, it is because we are biological entities, with an evolutionary past, that we have certain needs and propensities which lead at least many of us, in common circumstances, to have a recognisably moral concern for the well-being of the natural world. We also have propensities and needs which lead us in the opposite direction – towards destructive tendencies and heedlessness with respect to the other beings with which we share our world. In his writings on environmental ethics Wilson seeks explicitly to strengthen the former tendencies by dint of argument and persuasive rhetoric, and to reduce the appeal of the latter.

It will be important to discover whether what he says about the rooted tendencies he ascertains is actually true, or at least plausible, and if so, whether the prescriptions he offers for the implementation of an environmental ethic are realistic and effective ways to attain the environmentally ethical goals he postulates. Some sociobiologically-inclined thinkers (such as Ridley 1996) have been sceptical about whether we have it in us to accept and implement an extensive environmental ethic. Wilson is cautiously optimistic.

The works in which he has most extensively explored and defended an environmental ethic are, in the main, ones in which he spends considerable effort in elucidating the biological history of the planet, as a contribution to our understanding of precisely what is at stake in the development of such an ethic. This he does wonderfully well in such works as *In Search of Nature*, *The Diversity of Life*, *Biophilia* and *The Future of Life*. His overriding aim in the course of the elucidation of the history of life on Earth is to prepare the case for the protection of biodiversity as the key aim of environmental ethics.

Let us see how such arguments are conducted by examining closely his most recent work in this genre – *The Future of Life* (2002).

It will first be useful to take note of the general view that Wilson offers in this book of how the human moral sense develops. He suggests that 'ethics evolve through discrete steps, from self-image to purpose to value to ethical precepts to moral reasoning' (Wilson 2002: p. 131). The self-image mentioned here is both personal and social, or perhaps personal as mediated via the social, given our inherently social nature. Whatever this means precisely, it suggests that for Wilson, humanity's moral sense develops from within human beings, emerging as part and parcel of the development by human beings, differently in different cultural and other circumstances, of a basic conception of who they are and what, given who they are, are their important purposes. As we have just seen, this is the perspective on morality which is characteristic of naturalism, which views morality as a product of the contingent nature of a specific set of beings at a particular time and place.

Wilson, in this book and elsewhere, makes many attempts to ground human moral thought and behaviour in claims about the nature of the human self from which morality is generated. There are, as we discovered in our earlier survey of his case for gene-culture co-evolution, features of the human self which he thinks can be seen to have an evolutionary origin, and thus to have a ubiquitous influence in human morality, whatever the particular form of cultural mediation which is postulated.

We can once again see this pattern of thought operating right from the start of *The Future of Life*. In the opening chapter of this work Wilson addresses Thoreau as someone who states clearly the first element in a desirable environmental ethic – what Wilson here refers to, borrowing Aldo Leopold's phrase, as a 'land ethic'. This is to see the natural world as a logical place to turn, as Thoreau turned to it in his sojourn in Concord, in order to 'search for wholeness and richness of experience' (Wilson 2002: p. xxi) when everyday life is filled with trivial but unavoidable distractions. The urge to find such wholeness and richness is regarded by Wilson as part of human nature, as too is the 'human proclivity to embrace the natural world' (Wilson 2002: p. xxi). He credits Thoreau with 'hitting on an ethic with a solid feel to it ... nature is ours to explore forever; it is our crucible and refuge; it is our natural home ... save it, you said: in wilderness is the preservation of the world' (Wilson 2002: p. xxii). He expands on this point elsewhere:

> Into wilderness people travel in search of new life and wonder ... Wilderness settles peace on the soul because it needs no help; it is beyond human contrivance. Wilderness is a metaphor of unlimited opportunity, rising from the time when humanity spread across the world ... godstruck, firm in the belief that virgin land went on forever past the horizon. (Wilson 1992: pp. 334–5)

The view that the natural world, especially that portion untouched by human hand – wilderness – is a key resource for human moral and spiritual regeneration certainly appears to provide a firm basis upon which one might establish an environmental ethic. Also, the fact that some human beings find this a compelling approach to nature is undeniable. But the idea that there is a universal, or at least very widespread, human trait embodied in this approach is clearly highly speculative.

This claim, that the natural world is a logical place for human beings to turn in search of wholeness and richness of experience, as suggested by Wilson's reading of Thoreau, is distinguished from a related proclivity of the human mind to which Wilson often has resort in seeking to ground an environmental ethic – what he refers to as 'biophilia' (Wilson 1992 [1884]: pp. 333–5), defined as 'the connections which human beings subconsciously seek with the rest of life' (Wilson 1992: p. 334). This encompasses various phenomena.

Firstly, it is evidenced by the phobias experienced by many people of hazards which beset our ancestors in their natural environment, such as heights, closed spaces (from which escape would be difficult), open spaces (from which threats could emerge at any point), spiders and snakes (often venomous) and so forth. Contrariwise, human beings rarely have phobias directed towards recently-devised or encountered phenomena, such as automobiles, electric sockets, guns, and so forth, that, objectively-speaking, pose much more of a threat to human well-being than do poisonous snakes and spiders. The instinctive fear of snakes in particular is one which we share with our primate cousins, but has come to acquire a considerable cultural overlay in the human case, with some cultures transforming the power of the beast into a positive as well as a negative force.

Secondly, it is evidenced by the tendency of most human beings to seek out certain kinds of landscape in which to live, namely 'a prominence near water from which parkland can be viewed' (Wilson 1992: p. 334). This is said to resemble the ancestral savannah country in Eastern Africa in which human beings originally evolved, and which provided the human species with its most favourable mixture of desirable landscape features – namely food, water and security. Thirdly, a great many human beings, when they have the leisure to do so, have a strong tendency to seek out rewarding interactions with the natural world in the form of camping, fishing, birdwatching, creating gardens and so forth.

Biophilia so characterised has the advantage over Thoreau-style 'moral and spiritual regeneration' of being easier to relate to evolutionary processes and of being easier to specify. However, although it is plausible to explain it by reference to a hard-wired tendency in the human mind, it underpins a negative as much as a positive attitude to the limited sets of natural phenomena which are its object. At best it seems to underpin a positive attitude towards certain naturally-occurring kinds of landscape, an attitude which may lead human beings to preserve those kinds. But it also seems to underpin an environmentally-harmful propensity to alter other kinds of landscape so as to resemble the preferred one, thereby destroying habitats and ecosystems of immense value to human and other forms of life. This may be part of the motivation for the widespread human tendency to turn dense forest into savannah-style parkland. The phobias cited may make it hard to generate much sympathy in many human beings for the conservation of threatened species of snakes and spiders. Fishing, gardening and other highly managed forms of interaction with nature may lead to the desire to protect some species and habitats involved in such activities, but may lead to the destruction or neglect of species which do not enter into the favoured activity. Biophilia, therefore, is a highly ambiguous proclivity when viewed from the perspective of environmental ethics.

Leading on from these speculations about the 'spiritual regeneration' effects of wilderness, and the presence of biophilia, Wilson adduces a further reason for human beings to take an interest in nature, which takes us in a more social direction. This involves claiming that we do in fact care for at least some other species which have, for reasons 'difficult to understand and express ... become part of our culture' (Wilson 2002: p. 105). There is undoubted truth to this claim. Some species are ones which humans within certain societies have come to know well, and to which they have attached certain kinds of symbolic and other value. Their loss, or threatened loss, is often the occasion of special alarm and sadness, at least for significant numbers of members of a culture or society, though rarely all of them.

Wilson does not say so, but it might be that we human beings are hard-wired to invest at least some parts of the natural world with cultural significance. This would be unsurprising, insofar as human beings have a motive to get to know at least some aspects of that world very well, to understand and attempt to control it through measures designed to identify with it, interact with it and placate it. Of course, we shall then expect different cultures to identify with different elements of the non-human world. A universal trait need not be put into effect in the same way everywhere.

This trait, then, if it exists, is highly enculturated. But cultures rise and fall, change extensively for a whole variety of reasons, and are seldom entirely self-consistent even when at the height of their influence. Such traits may enable proponents of biodiversity protection to gain an initial hearing for their proposals, when directed to the culturally favoured species under threat. But the culturally favoured species may not be under threat, or their significance within the culture may be transient, superficial and/or localised. This phenomenon, then, may be another problematic basis for a viable environmental ethic.

It does, however, share the feature with the other putative bases of an environmental ethic so far mentioned, of being anthropocentric. Biophilia is a purely anthropocentric value postulate. So too is Thoreau's claim for the moral and spiritual value of wilderness. Nature and wilderness have value for a certain kind of creature – an intelligent mammal with an evolutionarily produced tendency to 'pursue personal ends through cooperation' (Wilson 2002: p. xxi), thereby acquiring a moral nature, albeit one which, as we have seen, is weighted by evolutionary forces to aim in certain directions – self, family, others, in that order. This is clearly not immediately a basis for an environmental ethic that postulates the intrinsic value of nature. It might be made into such an ethic indirectly, if it could be shown that the attribution of intrinsic value to non-human nature is an indispensable prerequisite for nature to have its beneficial function of giving human beings, in certain circumstances, a sense of wholeness and richness in their lives. But that is an argument which Wilson does not himself pursue.

Indeed, most of the time Wilson's environmental ethics are expressly anthropocentric. It might be thought that this an inevitable outcome of a view of the emergence of morality among human beings which sees it as having an adaptive function – that it helps to produce their inclusive fitness. It might be urged that a sociobiological approach to the understanding of human morality can only ever produce a strongly anthropocentric environmental ethics. But it is not obvious why it should suffer this restriction. As we have noted elsewhere, it is only if altruism is genuine that morality works in such a way as to benefit the moral being. But once altruism of this kind has been established as a tendency in the human mind then it seems eminently possible that it will be open to taking the interests of others into account – whatever species those others belong to.

It is important to note that Wilson does sometimes accept that there are defences of a moral approach to the non-human world that are expressly non-anthropocentric (Wilson 2002: pp. 133–4). But this is a view that he neither rejects nor embraces. This suggests that he sees that there are no reasons of evolutionary principle for rejecting the coherence of such a position, but that he thinks that the anthropocentric alternatives are more compelling and/ or likely to be more persuasive.

Sometimes, too, when he seems to endorse the intrinsic value of non-human organisms, he elides this into an instrumental, and so anthropocentric, position. For example, he tells us that

> [t]he creature at your feet dismissed as a bug or a weed is a creation in and of itself
> ... The ethical value substantiated by close examination of its biology is that the life

forms around us are too old, too complex, and potentially too useful to be carelessly discarded. (Wilson 2002: p. 131)

Here we move smoothly from characteristics of organisms that seem intrinsic to them – their age and complexity – to the non-intrinsic, contingent, and instrumental value that they may have for human beings.

We are not told why age and complexity should be properties that ground the moral value of the non-human. In line with his general position on the development of moral thought mentioned earlier, Wilson ought to try to show how the aim of preserving organisms which are ancient and complex emerges from some human self-image which, for preference, has a universal, because evolutionarily-produced, character. But he does not seek to do this, and thus his argument, even in his own terms, is left unsupported by his more fundamental moral theory.

However, arguably he does attempt this in a related argument that is based on the point that all life shares a common descent, and thus is genetically unified (Wilson 2002: pp. 131–2). This fact, when we recognise it, enables us to come to a certain view of ourselves which has moral importance, namely that we view ourselves as the conscious and morally-responsible part of the complex whole of life on this planet. In addition, the story of evolution satisfies the human need for a narrative which places us in a meaningful context and elucidates our purpose in being here – namely to act as the morally aware stewards for the rest of life. (Wilson 2002: pp. 132–3). Wilson at this point even goes on to posit yet another proclivity of human nature – stewardship itself, which, he tells us, 'appears to arise from emotions programmed in the very genes of human social behaviour' (Wilson 2002: p. 132). As often is the case with these kinds of claim, it is unclear upon what evidence it is based. This argument does, however, try to elucidate how a sense of the moral standing of the non-human world might be connected directly to a certain sense of self.

Yet, whatever its basis, this argument, although appealing, has obvious flaws as a basis for attributing intrinsic value to the non-human. It is not clear why the fact that human beings share descent with all other life forms should make the latter morally considerable for the former. Contrariwise, if Darwinism should turn out false, and it could be shown that some other life forms do not share a descent with us, it is unclear why that in itself should mean that they could not be morally-considerable. For many human beings, the fact that we share a common descent with other life forms is no more significant, morally speaking, than is the fact that all human beings share such a descent. These facts may, intelligibly, occasion nothing more than a shrug of the shoulders and a 'So what?' reaction. The story of evolution may be interpreted in the way Wilson suggests, but apparently the meaning contained therein for human life is, for many human beings, an inadequate one. As far as they are concerned it does not make us central enough to the story, and, importantly, it contains no account of what happens beyond this life.

The most we can get from these arguments, then – and perhaps the most that can reasonably be asked of them – is that they provide an account of a sense of self that may appeal to some, perhaps to many, individuals in at least some cultures and that may be used to support the case for moral stewardship with respect to the non-human world. But they do not appear to have a compelling appeal for all human beings, even given enough time for them to be propagated and even granted the claims Wilson makes about the rootedness of the desire for stewardship and narrative in human beings' universal nature.

To return to Wilson's discussion in *The Future of Life*, he goes on to suggest that 'the mood of Western civilization is Abrahamic' (Wilson 2002: p. xxii), by which he means that the natural world is conceived of in that civilisation as provided (by God) as an endless resource for human well-being. This claim partially deprives the argument from cultural significance of some of its force, by highlighting that the significance that the natural world may have for a culture may in important ways be environmentally harmful. Clearly, in suggesting that the Abrahamic mood is a cultural rather than a biological phenomenon, Wilson leaves room for the possibility that there might be civilisations which did not share that mood – and that perhaps even Western civilisation might be induced to move away from it. But this point does not detract from the fragility of any argument for moral concern for the non-human world that rests primarily upon cultural factors.

However, it is not really the idea of the natural world as an endless resource that he is concerned to combat, but rather a too insensitive and unnuanced version of what it is a resource for. Thus, the value of Thoreau's contribution is that he has found an important concept of 'resource' to apply to untouched nature – the spiritual. The preservation of this resource actually requires the conservation of wilderness, which is why it is a concept of resource that might serve usefully as an element in the argument for the preservation of biodiversity.

The aim of Wilson's environmental ethic is, thus, best thought of as being to find arguments for the preservation of biodiversity that build upon the various tendencies in the human mind that he has noted – the tendency to find spiritual regeneration in wilderness, which Thoreau highlighted; biophilia; the tendency to find cultural significance in nature; the tendency to be attracted to stewardship – and which take us in the direction of caring for the natural world. On this basis we can then seek to find ways of counteracting the Abrahamic mood which appears to be so dominant in Western civilisation – though he is well aware that strong elements of that mood are prevalent in other contemporary 'civilisations', such as the Chinese.

However, there is not simply what one might call a general moral concern operating in Wilson's discussion. He is not concerned to get the ethics right for its own sake. Rather, as is the case with all contemporary environmental ethicists, his discussion is motivated by the need to deal with what is perceived by him as an impending, or perhaps actually existing, environmental crisis, specifically with respect to biodiversity. This crisis he refers to as the 'bottleneck'. By this he means the need to enable a large human population, that he hopes will soon be stabilising in the region of eight to ten billion people, to escape poverty in such a manner as to permit the continuation of other species in numbers and habitats that will enable those species to remain in existence for the indefinite future. This crisis perspective may well be what contributes to the almost exclusively anthropocentic character of Wilson's approach. For time is pressing, and it may seem a wiser move to appeal solely to human needs and interests, rather than spend time trying to gain acceptance for an ethical perspective which may be hard to understand and unpopular when it is understood.

Wilson's discussion of Thoreau ends with some rhetorical flourishes rather than arguments: 'Surely the rest of life matters. Surely our stewardship is its only hope' (Wilson 2002: p. xxiii). He seems to suggest at this point that ethical propositions and value judgements come from the 'heart': 'We will be wise to listen carefully to the heart, then act with rational intention and all the tools we can gather and bring to bear' (Wilson 2002: p. xxiii). But it remains somewhat unclear what the role of the 'heart' is in this thought, unless it is a shorthand term for the four proclivities already noted.

This unclarity over the precise nature of the sources of human moral motivation becomes especially problematic later on in his discussion, when he offers a view of human beings as hard-wired by evolution to be short-sighted:

> The relative indifference to the environment springs ... from deep within human nature. The human brain evidently committed itself emotionally only to a small piece of geography, a limited band of kinsmen and two or three generations into the future ... Why do we think in this short-sighted way? The reason is simple: it is a hard-wired part of our Palaeolithic heritage. (Wilson 2002: p. 40)

He does go on to counter-pose this hard-wired short-termism with an ability to think long-term. However, this latter ability is seen as mainly involving a theoretical rather than a 'felt' or 'common sense' perspective. The implication of this is that a universal environmental ethic will be something for which it will be very difficult to gain acceptance, but which is nevertheless necessary to guide human beings through the bottleneck already described.

Wilson does not cite the evidential basis for his claims of the Pleistocene origins of our hard-wired short-termism. He also does not account for the existence of the ability to think in long-term, universal ways. The existence of the latter ability is not in question, of course, since his own work is a prime exemplar of it. But the nature and origins of short-termism could do with some further elaboration, given the importance that it has for his discussion. Counterposed to the short-termism are, then, two forces pushing in the opposite direction – the 'heart' with its capacity to empathise with nature and seek personal wholeness within wilderness, and the capacity to adopt universal and long-term theoretical perspectives. The latter capacity can both spell out a viable universal environmental ethic, and tell the part of our reasoning faculty which deals with technical and scientific modes of thought how to put it into practice, so as to steer us through the bottleneck.

For all that Wilson has said, however, hard-wired short-termism may be a dominant and powerful force in most human beings, and the feeling for nature and the ability to adopt long-term, theoretical perspectives may both be weak and too rare to have much chance of counteracting it. It may, therefore, be that it is some awareness of the problematic nature of the human psyche as a basis for long term, universal ethical thought, rather than just the crisis scenario, that accounts for the anthropocentric and heavily prudential nature of the specific prescriptions to secure the preservation of biodiversity which Wilson goes on the offer in the rest of his book and elsewhere.

Let us, therefore, now consider what Wilson has to say about the specific causes of anthropogenic biodiversity destruction and his prescriptions for its avoidance and repair in order to learn more about his view of the practicality of a universal environmental ethic.

The first point which Wilson is concerned to make is that it is biologically untenable to claim, of the many species currently declining, that they are doing so because species have a natural life-span just like individual organisms. As Wilson tells us, '[t]he great majority of rare and declining species are composed of young, healthy, individuals. They just need room and time to grow and reproduce that human activity has denied them' (Wilson 2002: p. 83). Human depredations of other species seem, however, to have had a long history. As Wilson documents, once human beings left their original haunts in Africa and colonised the other continents, they seem to have had a deleterious effect on other species, especially the megafauna that they encountered. However, this phenomenon can be explained in terms other than the attribution to human beings of a vicious killer instinct. As Wilson explains (Wilson 2002: p. 98), in Africa and tropical Asia, where human beings originated, they co-evolved with the surrounding megafauna, which had thus time to adjust to the human presence. Humans in any case at that period were insufficiently numerous and too underdeveloped technologically to pose too serious a threat to other species.

However, once they left Africa and tropical Asia, human beings encountered other species that were not used to their presence, and did not perceive them as a threat. Human beings were in effect introducing themselves as alien species into new ecosystems and had the devastating effect on other species in those ecosystems that alien species – such as rats on hitherto isolated islands – often have.

This explanation, however, although it does not cast human beings as genetically murderous with respect to other species also implies that we have not been subject to any natural pressures in the environment of evolutionary adaptation to make us environmentally careful. In the African/tropical Asian context we could do little harm because of co-evolution. There would be no particular need to be solicitous of other species' well-being. By the time we moved into the more ecologically vulnerable regions, where we began to wreak havoc on competitor and prey species, the lack of environmental inhibitions was part of the human norm. This is another aspect of the hard-wired short-termism that we have seen Wilson to postulate. The only counterforces to this, as we have seen, are the four postulated proclivities – biophilia, Thoreauesque spirituality, cultural significance of non-humans and the stewardship tendency – and the rational faculty for taking the long-term view.

The problem with this, of course, is that the first four, for the reasons already noted, do not seem up to the task of grounding any very extensive environmental ethic, and the latter imports into the discussion a practice of purely rational thinking which is inherently ambiguous with respect to the protection of biodiversity – or any other environmentally-protective purpose. This can be seen when we consider the rationally defended prescriptions which Wilson offers. He calculates that if we begin now to launch a concerted rescue package for the threatened biodiversity hotspots of the world it should be possible to cut the projected loss of species in those areas 'by at least half' (Wilson 2002: p. 102) by the end of the twenty-first century. This means, however, that, on the basis of Wilson's own calculations, there will still be a loss of some 25 per cent of the species in those hotspot areas. From the point of view of purely rational considerations, and even accepting that Wilson's projections are accurate, it may well be argued that this is still too large a rate of loss to be worth all the effort to attain. Why bother with an 'all-out' effort if extinction rates remain so high? Of course, much more rational argument can be given for seeking to secure this outcome, as we will be noting shortly. But once a course of action is viewed as appropriately to be determined by purely rational calculations the whole matter begins to turn on a complex interplay of judgements and interpretations of arguments, data and prognostications, even assuming agreement on fundamental value positions.

In any case, Wilson himself gives us further reasons for pessimism about our rational capacities for altering our behaviour in ways that will enhance and preserve biodiversity, rather than destroying it.

Firstly, he posits yet another human proclivity that takes us away from a concern for the non-human. This is our tendency for distraction and self-absorption (Wilson 2002: p. 105). Concerned as we are so overwhelmingly for ourselves, we tend not to notice precisely what are the effects of our actions upon other species. This tendency is not the same as the short-termism which we have already taken notice of, but it looks to come from the same stable, for it is said to be in 'our nature' (Wilson 2002: p. 105).

It also could plausibly be thought to subsume the other four tendencies that Wilson has categorised as supporting some concern for the natural world. After all, biophilia, spiritual absorption in nature, the finding of cultural significance within it and the tendency to stewardship can all be thought of as primarily directed towards the human self or selves. If so, then it will be a delicate matter to determine whether it is the directedness towards self, or the directedness towards nature located within that overall self-directedness, which is the more significant factor. It is this thought which makes certain versions of postmodernism seem so problematic to at least some environmental ethicists, for in those versions a concern for nature is simply a version of concern for the human self, and nature has no significance except in relation to human selves (Gare 1995).

Secondly, Wilson spells out a further implication of the point that we have not co-evolved with most of the species and ecosystems we are currently threatening. We are, he tells us, engineers of ecosystems, but we are not at all good at it: 'Not having co-evolved with the majority of life forms we now encounter, we eliminate far more than niches than we create' (Wilson 2002: p. 112). Hence, beneficial stewardship of the planet is not something we are naturally good at. We should not, either, console ourselves with the undeniably true claim that we can create, and have created, environments which benefit some other creatures greatly, for we have destroyed more than we have created. To be good stewards we will have to apply ourselves diligently to the task, learning all we can about an immensely complex set of tasks. This will require a concerted desire to do so.

In spite of these two considerations which appear to raise further barriers to human beings' taking effective care of the non-human environment it may be that the balance of considerations which a rational intellect would find worthy of note will sway the argument decisively in the direction of biodiversity preservation. As we can expect from the general logic of Wilson's position, in which the non-rational proclivities are partial, ambiguous and problematic, and the rational ones are complex and hard to settle definitively, the considerations that Wilson provides are yet again strongly anthropocentric.

We are thus given a cluster of arguments that focus upon the potential contributions to human material well-being to be derived from other species. He presents the case for the medical and pharmacological benefits to be realised from the exploitation of other species' genetic and chemical characteristics. In this, he emphasises the importance of serendipitous discoveries, and thus the importance of keeping as many other species in existence as possible. This then leads on to an economic case for encouraging the activity of bio-prospecting in hotspot areas, which potentially provides income and occupations for some of the poorest human societies on the planet (Wilson 2002: pp. 118–20).

Towards the end of the book we are presented with 12 key practical steps which Wilson urges us to take in order the secure biodiversity protection through the phase of the bottleneck (Wilson 2002: pp. 161–4). Most of these are practical measures which derive from the scientific understanding of the causes of biodiversity loss, such as the imperative need to protect the planet's 25 bodiversity hotspots which together cover only 1.5 per cent of the Earth's surface.

However, when these steps involve reliance upon the cooperation of human beings who may be indifferent towards, or ignorant of the importance of, biodiversity protection, Wilson once again proposes measures which are strongly anthropocentric. Thus, logging of old-growth forest should cease everywhere, but be replaced with tree-farming on already converted land; conservation should be made profitable, via such measures as ecotourism, bioprospecting and carbon-credit trading, for those living in or near to protected areas; biotechnology should be used to develop new crops and livestock so as to meet human needs while reducing environmental impacts.

The question of how these measures are to be undertaken is answered by Wilson in terms of the contributions by, and mutual cooperation of, governments, private corporations, scientists and NGOs. It is an optimistic account which posits a coming together of some groups which are normally ideologically hostile to each other in an effort to search for common ground (Wilson 2002: pp. 152–5). It also imports a set of assumptions from the consilience agenda. He tells us that:

> Within several decades, many neuroscientists believe, we will have a much firmer grasp of the biological sources of the mind and behaviour. That in turn will provide the basis for a more solid social science, and a better capacity to anticipate and step away from political and economic disasters. (Wilson 2002: p. 156)

This, of course, assumes that economic and political disasters have their origins in some aspect or other of the individual human mind, rather than in something to do with the workings of impersonal and malfunctioning systems that may be created by those minds, but that may have properties not explicable in terms of how those minds work.

This point leads on to a more general one connected with Wilson's analysis, which is that there seems to be something of a gap, between the ethical theory on the one hand and the practical outcomes on the other, where politics ought to be. It is clear that he believes that human beings have now started to converge on their basic economic and political commitments, and his own are basically liberal and pragmatic. This may seem to imply that there are no longer any serious questions to be addressed about the political organisation of human societies, as opposed to the particular policies that should be implemented by those societies.

But, even if the whole world were tomorrow to become unquestioningly committed to liberal norms and practices, there would still be serious issues concerning the most efficacious form of economic and political organisation within those liberal norms which could make a serious difference to the outcomes of Wilson's prescriptions. For example, it would be very useful to know how we should best cope with the short-termism of the liberal democratic system, which is a separate issue from, although it may be directly connected to, the short-termism which Wilson has already diagnosed as an evolutionarily-produced human trait.

There are also the key issues of how compromises should be reached between competing individual and group interests and values. Such competition works best when it is only marginal benefits and harms which are at stake, rather than the more far-reaching alterations in the pattern of rewards which may be consequent upon attempts to deal with the bottleneck Wilson has identified. It also works best when the cake of wealth is increasing in size. The problem here is that, as Wilson himself clearly acknowledges, the concept of the 'ecological footprint' emphasises the finitude of the size of the global 'cake'.

To put the point in the terms outlined by Wilson, the impact of specific societies upon the planetary environment is a function of their population size (P) times per capita consumption, or affluence (A), times the resource demands of their consumption technology (T). This PAT formula converts into a measure of the hectares of productive land needed to support the lifestyle of average members of the society, given existing technology and consumption patterns. This hectare measure is the 'ecological footprint' of the average member of the society in question.

Proponents of this conception, such as Wilson, multiply the footprint of average members of the most affluent societies on the planet by the current and projected world population. For example, the ecological footprint of a US citizen is 5 hectares, according to Wilson. If the current world population of 6 billion people were each to use this amount of resources the planet would need to provide 30 billion hectares of productive land. But this is not available on this planet, and thus, he tells us, we will need two more planet Earths to raise everyone to the US levels of affluence (Wilson 1998: pp. 314–15). Since we do not have two more planets, we will need to meet the legitimate needs of the poorest citizens of the world by sharing more equitably the resources that we do have. Issues of distribution become extremely pressing in these circumstances, both within and between societies. These are clearly issues with which our liberal democratic political systems may have great difficulty, since their electoral cycles encourage political parties to focus on the near future and to compete for votes on the basis that they are better able than their opponents to offer their populations greater affluence for them and their descendants.

Of course, the challenge of the ecological footprint may be met by the claim that technology can advance so as to enable human beings to maintain or achieve high consumption levels via the more efficient exploitation of our existing resources and via the discovery of new forms of resource. On this view, provided we have sufficient capital and ingenuity there are no practical limits to the attainment of high standards of living for all the planet's human inhabitants. Free markets and free trade are held to be the *sine qua non* of these developments. Wilson is sceptical of these claims, particularly on the basis of their lack of ecological realism. The problem is not just the finiteness of the planet's base of productive resources, it is also the finiteness of the planet's capacity to absorb the waste products of the productive activity, and the potentially catastrophic effects of ensuing large-scale ecological changes, such as climate change produced by human emissions of greenhouse gases (Wilson 1998: pp. 317–26).

Clearly, we are here reaching the key issues concerning the environmental future of the planet over which there is enormous controversy precisely because the scientific research upon which predictions are based is open to dispute. It may be that Wilson and the environmental alarmists are wrong, and that we will be able to secure high living standards comparable to those of the current average US citizen without seriously harming the planet's biosphere and the life-support systems upon which we all rely. There appears to be nothing in the Darwinian worldview to help us resolve these disputes. Proponents of that worldview may well be found on both sides of the issue, although, given that prominent exponents of the view tend to be biologically-trained and ecologically well-informed, the sensitivity to the dangers highlighted by Wilson tends to be more pronounced amongst them than one would expect amongst, say, social scientists who are trained to focus almost exclusively upon the human species in complete abstraction from its environmental and ecological interconnections.

Perhaps a sociobiological perspective upon the human mind may help us to understand these difficulties. But it may not. It all depends upon what kind of difficulties they are. The phenomenon known as the collective action problem, which bedevils environmental issues wherever we look, seems to rest on the capacity for rationality that Wilson relies upon as a key resource to get us through the bottleneck (see Connelly and Smith 1999: pp. 106–7). It also involves the distinction between individual self-interest and the common good that Wilson himself relies upon, as we have seen with respect to the anthropocentric suggestions contained in his 12-point plan of action. The collective action problem arises when we make what appears to be a reasonable assumption, namely that most people are not idealists with respect to environmental concerns, and thus are not prepared to change their behaviour in more environmentally-beneficial ways without attending to the balance of costs and benefits for themselves. It can then be shown that the rationally self-interested calculation about costs and benefits is that the cost to the individuals in terms of time, money and effort to alter their behaviour is usually much greater than the shared environmental benefits which the alteration in question will achieve. Hence, even when all can see that it is in the interests of all to achieve the environmental benefit in question, it is not in the rational self-interest of each to play their part in achieving it. Each realises this, and so nobody changes their behaviour.

There are solutions to these difficulties, which involve altering the patterns of costs and benefits to the individual – providing rewards to individuals that outweigh the costs, or penalising them for not altering their behaviour. It would be very useful to know if there are any prospects of the sociobiological perspective revealing the best way to deal with collective action problems, on the basis of its understanding of the bases of human behaviour and motivation. Perhaps the sensitivity of human beings to cheating by others, attested to by the findings of evolutionary psychology, could be cited as a reason for supposing that, once a regime of penalisation had been established, people could be relied upon to monitor cheating in the form of free-riding (Cosmides and Tooby 1992: pp. 181–206).

Conclusion

The upshot of this investigation of the case that Wilson makes for the development and implementation of an environmental ethic is, perhaps predictably, somewhat mixed. We can divide the matter into several problem areas. Firstly, there is the all-important issue of moral motivation. This covers the question of whether or not human beings have any natural or gene-based tendencies to be concerned to alter their behaviour so as to achieve environmentally beneficial outcomes in various circumstances.

As we have seen, the conclusions offered by Wilson on this topic are only somewhat supportive of the idea that environmental ethics can rely upon some epigenetically based human tendencies, such as biophilia and the attribution of cultural value to certain natural phenomena – places, species and so forth. But, in the main, these are partial, ambiguous and probably not fully universal traits. The faculty of reason can, and indeed must, play a role in developing moral concern towards environmental phenomena if such ethics is to have any practical import, but rationality is also a two-edged sword, and, as the collective action problem suggests, may itself be the source of impediments to effective implementation of the requirements of environmental ethics.

One possibility worth considering is whether these sources of human moral motivation with respect to the environment are strengthened by the adoption of the Darwinian perspective itself. Does the acceptance of the belief that human beings are an animal species – a species of primate – sharing common descent with all other life-forms, and formed by the process of natural selection have any tendency itself to promote the acceptance of an environmental ethic? Acceptance of this belief ought to heighten the sense we have of our interconnectedness with other life-forms, and of a common reliance upon a shared biosphere. This ought in turn to stimulate an interest in understanding other species and the common environment we all share. From this interest in, and understanding of, such species and habitats it ought to be possible to acquire a greater degree of concern for their well-being.

This movement of thought, however, stemming as it does from the acceptance of a purely intellectual, because scientific, form of understanding which human beings have only recently acquired, and which is accepted, or even known about, only by a relative few, cannot have any epigenetic basis. Acceptance of it might support those tendencies within human beings for which a plausible epigenetic case can be made, such as biophilia, which might in turn bolster environmental ethical concern. But it is not obvious that this is an automatic, or perhaps even very frequent process. Still, if there is an epigenetic tendency for most human beings to accept culturgens which imbue natural phenomena with cultural significance, as Wilson has suggested, than it is not impossible that the culturgens embodied in the Darwinian worldview can contribute to the granting of high cultural significance to nature viewed in the extensive sense characteristic of Darwinism.

For this to have much effect on human ethical motivation, however, the Darwinian worldview needs, once again, to achieve widespread acceptance. We have already noted this requirement in connection with the full achievement of the program of consilience. It now looks as if the development of an extensive environmental ethic amongst enough human beings to make a difference may also need widespread acceptance of the worldview itself, or at least of key components of it. In the next chapter we will be examining some of the basic problems which many have with the acceptance of that worldview with an eye to seeing how possible it is to meet their criticisms.

However, probably such widespread acceptance of the Darwinian worldview is at best a necessary, not a sufficient, condition for the development of strong environmental ethical motivation in human beings. After all, it is not hard to find professionally trained biologists, fully accepting of the Darwinian account of evolution, who have a completely instrumental attitude to the non-human world of nature.

There is, however, potentially another way in which the Darwinian worldview itself can contribute to the development of ethical motivation, and this is via its deployment in the intellectual case for the moral standing of the non-human. As we have seen, Wilson, although he sometimes gestures towards the idea that non-human nature has moral standing, nearly always offers moral arguments that are anthropocentric. He repeatedly tells us that we have responsibility with respect to the natural world, but only because of the ways in which it matters for human beings, albeit not in any narrowly practical sense of 'matters'. However, one of the repeated features of his arguments, and indeed of the arguments of Darwinians since Darwin himself, is to draw attention repeatedly to the ways in which human beings share features with other organisms with which they share common descent. This ought to heighten the sense that other organisms possess features, such as sentience, certain forms of intelligence, a sense of self, and so on, which ground a case for their moral status. They can then be said to possess at least some degree, and some kinds, of moral status in their own right. It then becomes possible to do wrong to them, rather than to do harm to other human beings with respect to them, by, for example destroying the animals that form part of some human being's livelihood.

Wilson never really presents or fully considers such arguments, although his own concern for the rest of the natural world for its own sake often appears in his writings. They represent by now a fully developed field of ethics in their own right. I have elsewhere developed a case for ecological justice on this basis, which is the idea that all other organisms possess a moral claim to a share of the planet's resources (Baxter 2004). The matter is too complex to be dealt with here. However, the key issue for this discussion is how far such moral arguments can be said to be supported by the Darwinian worldview. When taken together with the claims made by Wilson in support of the idea that we do have epigenetic tendencies to care about nature it can be fairly said that the connection is not negligible. Acceptance of the worldview can provide culturgens which, taken in conjunction with the epigenetic bases of moral action in general, and concern for the non-human world in particular, can provide a reasonably robust basis for environmental ethics.

However, of the two remaining elements in an environmental ethic – its practical prescriptions and the political dimension of implementation – the latter is particularly fraught with difficulties. Wilson provides some excellent prescriptions, but, as we have seen, does not have any very clear or convincing ideas about their political implementation. It has to be said that the difficulties here are not peculiar to him. No-one has produced any very convincing plan of political action for the attainment of environmental goals, except for those environmental aims which can most readily met within the existing international political framework. The most radical problems of all, the ones which Wilson and others use the idea of the ecological footprint to highlight, are the most difficult to articulate, defend against strong and determined criticism and make any show of solving by practical political means. We do not yet have, and perhaps we never can have, a settled politics of radical redistribution involving non-zero-sum games.

Insofar as sociobiological approaches have had much to say about the relevant political issues here they have arguably offered causes for pessimism. If human beings have, as Berkow hypothesises, an epigenetic tendency for hierarchy and nepotism, to seek to safeguard benefits for their offspring by forming alliances among powerful groups, then there will be very strong forces to overcome if global equity among human beings with respect to shares of environmental resources does turn out to require extensive redistribution.

Here, as elsewhere, we need some extensive and rapid research into the sociobiology of human motivation, which gives yet one more reason for asking the question of how far the Darwinian worldview is capable of acceptance on a large enough scale to make the kind of differences anticipated by those who make the case for sociobiology and environmental ethics. Let us, therefore, now consider some of the serious objections that many have to the whole Darwinian worldview.

Chapter 9

Evolution, Meaning, Suffering and Death

We have so far in this book been considering two main questions – whether the evolutionary perspective bequeathed to us by Darwin is able to cast much light on the nature of human beings, and whether it is able to sustain an account of morality which is convincing, illuminating and able to provide a clear and robust environmental ethic.

There is, however, an even more fundamental issue raised by Darwinism that we must now address. We need a discussion of the implications of Darwinism for the possibility of meaningfulness within human life. This is because for many the world according to Darwinism is too random, too much the subject of chance, too arbitrary in terms of who does well and who does not, to be anything other than deeply unsatisfactory, at least from the human point of view.

Further, the theory of evolution is not teleological. There is, Darwinists frequently emphasise, no underlying point or purpose which it is necessary to postulate in order to make evolution by natural selection an intelligible or defensible theory. Hence, the Darwinian worldview lacks for many the essential ingredient – an overarching purpose to life – which could make the possibility of a meaningful life for individuals a real one.

In partial mitigation of the non-teleological nature of Darwinism, some Darwinists claim that natural selection may produce a deepening complexity in the world of nature with the passage of time, and to that extent the process has some in-built tendencies which assert themselves given enough time (Dawkins 2004: pp. 609–18). But which kind of complexity this is, whether there is some point to nature's manifesting complexity of one kind rather than another, and whether it is a sufficiently strong tendency overall, are all questions which indicate the problematic status of the idea that this aspect of evolution can provide an overall sense of meaning or purpose to life.

Another key ingredient in Darwinism has been emphasised by Lisa Sideris in her recent discussion of ecological theology (Sideris 2003). She argues that the nettle of suffering and death must be grasped as an ineradicable part of the Darwinian worldview. Her target in saying this are those theologians who have held what she takes to be the too easy, and in any case factually incorrect, view that the theory of evolution supports the idea of nature as a harmonious whole. On this basis they maintain that the Darwinian perspective can be reconciled with the existence of a benevolent creator. However, she argues, the processes of evolution by natural selection are often violent, cruel and wasteful, at least to the human gaze. This is because the physical universe contains enormously strong forces which occasionally break down the ordered patterns which the laws of nature, of which those forces are a manifestation, have produced.

Such forces are no respecters of harmony and beauty, blind as they are to everything. Yet harmony and beauty themselves emerge by the very same processes, such as those of natural selection. Death and suffering are the very phenomena that shape and create many of the most wonderful manifestations of the natural world, including ourselves. The first issue to which we need to give our attention in this chapter is whether any of these thoughts justify the view that the Darwinian perspective helps us to come to terms with the reality of death and suffering at least as well as, or even better than, those positions which appeal to some notion of overall purpose behind the universe, such as the theodicy of traditional forms of theism.

Arguably, traditional theodicies are scarcely imposing examples of rigorous reasoning. Faced with the obvious fact that evil can visit even the most virtuous human being, an idea explored in the Book of Job, traditional theology is usually led to postulate at some point that there is a great mystery in God's purposes which our finite and limited intellects will never be able to fathom. We must await a revelation, in this world or the next, before we have hope of understanding how what appears to be random suffering is in reality part of an intelligible and morally satisfying plan. This looks to be an attempt to avoid giving an explanation, rather than furnishing one.

However, we should first take note that though we are here contrasting Darwinism with theodicy, and although Darwinism often goes together with an agnostic or atheistic viewpoint, some have taken the view that there is no necessary connection between the theory of evolution and atheism, or anti-religious positions more generally. Indeed, the Catholic Church has come to the view that it is possible to see natural selection as the means by which a deity has chosen to fashion the world, human beings included – at least, this seems to be the message of Pope John Paul II's letter 'Truth cannot contradict truth' sent in 1996 to the Pontifical Academy of Sciences (Christus Rex Information Service 2006). But this suggestion prompts the question of whether the accommodation of Darwin's view to the needs of theology, at least of the Judeo-Christian–Islamic tradition, is really possible, given the role of death and suffering, and the enormous influence of chance that, as we have just been noticing, is inherent in evolution as Darwinism describes it. Let us consider some possibilities that may be available to those who wish to reconcile theology and Darwinism.

To take the problem about the role of chance first, the difficulty is that for traditional theism the emergence of human beings, made in the image of God, has to be an intended part of the divine purpose. But the evolutionary account offered by the Darwinian theory of natural selection emphasises the important effects of random factors, including massive extinction episodes, throughout the history of life on this planet.

Perhaps one way of dealing with this problem is to suggest that evolution by natural selection, wherever it occurs, is bound to lead eventually to the emergence of intelligent life-forms of which human beings are an example. Perhaps God could know that he was on his way to producing such life-forms by this means, however the processes played out in detail, including the occasional global catastrophe. After all, the idea of human beings as being made 'in the image of God' seems to be a formula for alluding to their possession of a certain kind of moral intelligence, not a means of claiming that God has the appearance of a certain species of primate. Hence, it may just be happenstance that we look the way we do, but a matter of evolutionary inevitability that some creatures or other come to possess moral personhood – it just happens to be us, on this planet. However, a crucial problem with this proposed explanation is that there is no sound Darwinian reason for supposing there to be this putative inevitability. Theologians are still free to posit such inevitability for their theological reasons, but it does not appear to be supported by the Darwinism that it is the goal of this approach to integrate into the theology.

There is an alternative form of explanation of how God could have intentionally created human beings in his image, which seeks to respect the role of happenstance in the Darwinian account. This is to hypothesise that the apparently random catastrophes and other events, which seem to have been crucial in the chain of events which led to us, were in fact acts of God at crucial junctures in the evolutionary process. They should thus be thought of as creative interventions of God, turning the course of evolution in the desired direction.

The theological problem here is that this allows the possibility of God's intervening in natural processes. If God can do this if he chooses, then it becomes mysterious why He does not do so all the time, specifically to prevent the infliction of evil on the just and the wholly innocent, such as very young children. Here we touch on the second problem with the reconciliation of theology with Darwinism – the pervasiveness of unmerited suffering in the Darwinian processes, at least once sentient life has emerged. We are then presumably thrown back on the inscrutability of the divine purpose to deal with this problem. But appeal to such inscrutability, as we have already noted, looks like the refusal of an explanation rather than the provision of one.

A further theological possibility to cope with the problem of chance in the evolutionary process is that God sets the process in motion, waits for it to throw up the right kind of life-form to be the bearer of an immortal soul, and then, in unique act of divine intervention, inserts such souls into the animal carriers. Something like this seems to be the view preferred by the Catholic Church, as expressed in the letter by John Paul II referred to above. It gets round some of the difficulties already mentioned, but requires an explanation of why this is done at all, especially as the ultimate destiny of such souls is apparently to be released from any animal-like form. It also raises the empirical question of when and where this was done, and the morally acute question of what the point could be of inserting such souls into a context of randomised suffering.

The first difficulty, of the role of chance in the process of Divine creation, faces theology largely as the result of the attempt to reconcile it with Darwinism. The second problem, that of unmerited suffering, confronts theodicies even without the introduction of Darwinism into the discussion. For example, even if creationists are right, and the account given in the Book of Genesis is literally true, the problem of why the just, the innocent and those wholly devoid of moral agency, such as all other life-forms, suffer so much will need to be answered in such a way as to make such suffering intelligible and justified. The traditional answer which appears to be offered by the Book of Genesis, that such suffering is a punishment for an original act of transgression against God's express command, leads to a whole host of other puzzles as difficult to resolve as the problem this story was intended to address. If Adam and Eve were not moral agents prior to eating the apple, but only acquired the knowledge of right and wrong after eating the forbidden fruit, then it seems bizarre to punish them. What was the tree doing in the garden in the first place? Who let the serpent in? And so on.

However, if Darwinism is accepted as the sole truth about how life, including ourselves, came into existence then there is at least the prospect of demonstrating the intelligibility of such suffering, even if we have from this perspective no prospect of showing that it is justified. Since justification is the aim of theodicy, taking up this Darwinian standpoint means expressly forgoing the prospect of a satisfying theodicy. Instead, we have an explanation of the precise way in which death and destruction is an inevitable consequence of the laws of nature and of the workings of the blind natural processes which such laws enable to exist. There are good Darwinian reasons for the emergence of sentience, and thus of the possibility of suffering. There are other such good reasons for the emergence of a moral sense, self-awareness and the anguish and dread that these inevitably facilitate once the facts of existence are known to the human mind. There is, however, no good reason to suppose that blind natural forces will allocate pain and destruction in any morally-patterned manner, for such moral patterns emerge only from the activities of moral agents, of which human beings are the only known exemplar.

Could there not be a theodicy fashioned out of this Darwinian position to deal with the second problem? (The first problem would remain, of how to get round the role of the random in Darwinian processes so as to guarantee the emergence of human persons.) Perhaps it might be argued that God had no choice, if he was to create the world at all, than to create it in accordance with the physical and biological processes we observe, with random and unmerited suffering the result. The theodicy would then consist in the claim that these unmerited forms of suffering will be compensated for in a later life. It might even be added that the undergoing of undeserved suffering is an important element in the development of the individual's soul preparatory to that life. On this view one might even reinterpret the incarnation of Christ as an act of solidarity between God and his creation – demonstrating a Divine willingness to share suffering which not even God has the power to remove.

However, this would have the great drawback, at least from the viewpoint of mainstream Christian theology, that it would not quite fit the picture of Christ's suffering as an act of redemption, which fits more with the already noticed Genesis story of humanity bringing suffering on its own head as the result of an original act of disobedience, so that some intercessor is needed to take the burden of punishment away from us. If God had to create a world of random suffering, or not create a world at all, then the existence of such suffering cannot obviously simultaneously be viewed as a punishment. And, of course, a further theological problem with the claim that a world of random suffering is unavoidable for God is that he appears to have created a form of existence in which that form of suffering is conspicuously absent in creating the garden of Eden. That is why the account of the Fall is supposed to be so poignant – we can see what we have lost.

To return to the outline of the Darwinian perspective and what it has to say about the existence of unmerited suffering, we may note a further contrast which it manifests with characteristic religious positions. Religious believers who are theists are required to be grateful to God for his having bestowed life on them. But there is no reason from the Darwinian perspective for moral agents, such as human beings, to be grateful to the blind natural forces which have created them, for gratitude is only owed to those who intentionally benefit us. Many human beings, however, can intelligibly be glad that such forces exist, for we all do owe our existence to their workings, and if our lives contain a preponderance of good experiences over bad ones, then it is not unintelligible to be glad that we exist. But it is also not unintelligible for at least some human beings to wish that, on balance, they had not been brought into existence by such forces. Either point of view can be reasonable, from the Darwinian perspective.

Since we are moral beings, and many of the things which make human life unbearable stem from the acts and omissions of human beings towards each other, we arguably have a moral responsibility to try to ensure that the class of human beings with good reason to regret their existence is as small as possible – not by eliminating those human beings, of course, but by eliminating as far as possible the causes of their suffering, especially those for which we bear sole or main responsibility. In this endeavour, if Darwin's account is correct, we can not expect any particular help to come form natural forces, though equally we need not expect any specific hostility to our attempts in this regard either. The indifference of the natural world works in both directions.

To many, especially religious believers of all kinds, this account will seem to suffer from some obvious drawbacks. There is first the sheer ghastliness that many will find in the contemplation of a wholly non-purposeful, and amoral, universe. Then there is the problem that if suffering can fall upon the good irrespective of their goodness in a way that is not cancelled out by some future reward there is no obvious benefit from being good. Why doesn't Darwinism entail that the clear-sighted approach to life is to be as bad as we can get away with, for we may be good and suffer horribly, or bad and experience a life of endless pleasure?

Here Darwinism can reply in the manner of many of the moral teachers of antiquity. For human beings, virtue is a matter of being fully human, and being fully human is to live most truly by our natures. We have evolved as a species with a strong moral sense, and our lives have the most meaning, from the personal perspective, when we live most closely by that nature. In other words, the answer to the question of why we should be good is to be given in terms internal to human life, not by reference to some benefits delivered by means of some transcendent reality. This is the naturalistic perspective once again, focusing, as we noted earlier, on the phenomenon of integrity. It also has the benefit of coinciding with a defence of the moral life that has been articulated completely independently of the Darwinian perspective. Many moral thinkers have found it persuasive. If so, it should not suddenly become less persuasive just because it is supported by Darwinian reasoning.

The Sociobiological Account of Religion

We saw in an earlier chapter the ways in which morality is to be accounted for in a sociobiological approach, and the ways in which the naturalism that this approach embodied might be defended against the main criticisms of it. But religion also is a key ingredient in human behaviour, and requires to be explained in Darwinian terms. To an extent we have already touched upon key issues in this area. We have already noted some of the difficulties that stem from the attempt to reconcile Darwin's theory of evolution with traditional religious beliefs in the Judeo-Christian–Islamic tradition. We have compared religious theodicies with the Darwinian understanding of suffering and death. We need now to consider directly the implications of Darwinism for the tenability of religious belief.

In explaining the emergence of religious belief, does Darwinism inevitably end up explaining religion away? Or is it the case, as many still argue, that Darwinism, being a scientific theory of natural causation, cannot properly be regarded as in competition with religious beliefs, which aim to answer a completely different, but even more important set of questions – concerning point the or purpose of the world we inhabit? Or is it that Darwinism has rendered religion obsolete by revealing the evolutionary origins of our characteristically human, spiritual, search for meaning and purpose, thereby also revealing the only intelligible way of dealing with those questions, from within the world of experience, not by seeking to penetrate beyond it? Does Darwinian naturalism effectively eliminate transcendentalism in religion as well as in morality?

Let us first consider what Wilson has to say about religion in his book on consilience. His first point is to note that systems of religious belief are numerous and usually finite in their duration. Their life-cycle is dependent upon the needs of the human beings who create and support them. This implies that in order to understand religion we need to see it in functional terms.

Religions, then, he sees as beginning as essentially tribal phenomena, serving to cement a certain set of moral rules within a distinctive group of people by giving those rules a transcendental sanction, often embodied in a creation story which puts the people concerned at the centre of the universe (Wilson 1998: pp. 285–6). The evidence suggests, he explains, that moral thinking among human beings predates religious belief, and that religion developed as a means of giving force to accepted moral precepts among specific tribal communities, enjoining them to compete with rival tribe/religion complexes, and to sacrifice themselves in defence of their own religion/tribe.

However, Wilson is quick to explain that the roots of religious belief are not limited to this connection with the morality of the tribe. Other powerful epigenetic factors operate to bolster the religious drive of human beings. The survival instinct is one such powerful source – the strong desire to continue in the face of the omnipresence of death, for ourselves and all other living things (Wilson 1998: p. 286). Religions that hold out the promise of life after death will have strong appeal from this point of view. It has to be said, in passing, that not every religious view does promise this. Some, such as the original form of Judaism, did not (Midgley 2002: p. 130). Many, for example among traditional African religions, are primarily concerned with making this life go well. Even where religions do posit an afterlife, they sometimes, as in Graeco-Roman religion, depict a post-death existence that holds little attraction, even for the believers. In such cases, the orientation of the religion is rather towards patterns of required conduct, and fitting behaviour, within this life. This may suggest that the survival instinct element in religion may not be as important as that of supporting the appropriate moral dimension within this life. Or perhaps that the survival aspect only comes to the fore within certain cultural settings.

Another epigenetic source of the power of religion in human life is the characteristic human urge to achieve useful causal understanding of the threatening, but potentially beneficial, natural world around us: 'Doctrine draws on the same creative springs as science and the arts, its aim being the extraction of order from the mysteries of the material world' (Wilson 1998: pp. 286–7). Lewis Wolpert, another exponent of a Darwinian account of the epigenetic origins of religious belief, has recently offered a similar mode of explanation (Wolpert 2006). He cites the ability of human beings to grasp the notion of causality and put it to use in the development of increasingly complex forms of technology as the mainspring of religious belief. For, once human beings understand cause and effect, then they inevitably try to apply it to the fundamental, and distressing, facts of human life, such as our mortality, unmerited suffering, and the rest. They are led to posit explanations for these phenomena and seek ways of manipulating the world, via religious practice, to avoid, mitigate or seek consolation for them.

The final important epigenetic influence on the tendency for human beings at all times and places to develop some form of religious belief is tribal identity itself. In participating in religious belief sanctioned by the tribe one is simultaneously witnessing and participating in tribal life. There is enormous comfort and sustenance from this fact alone. One is validated in one's membership, one achieves dignity and recognition, and consolation and support at key moments of disturbing transition – the rites of passage: 'Others have gone here before you – do not be afraid, it is understood what has to be done in this situation, we are all with you and the gods themselves have told us what to do.'

Wilson notes that the emphasis which Darwinian naturalism puts upon the epigenetic bases of religious belief, a basis which he takes to be required in order to explain the ubiquity of such belief, does not in itself show that any particular elements in religious beliefs are untrue. Rather, what Darwinian approaches serve to reveal is that religious belief is based upon genetically-produced tendencies in the human mind. The tendency to such belief has an epigenetic basis, and thus during the period when such epigenetic bases were forming, the beliefs in question must have been of benefit to the inclusive fitness of members of tribal groups which held them: 'There is hereditary selective advantage to membership in a powerful group united by devout belief and purpose' (Wilson 1998: p. 287).

Wilson then presents the usual sociobiological calculus, derived from population genetics, as applied to the explanation of altruism of the most extreme sort, involving the sacrifice of the individual for others in whom the individual's genes are represented. Genes underpinning such altruism will spread through a group if the survival of the group is so enhanced that it outweighs the reduction in individual survival consequent on the tendency to altruism (Wilson 1998: p. 287). Hence, religious encouragement of self-sacrifice for the tribe and its gods can have a perfectly intelligible evolutionary basis.

From the point of view of human phenomenology, as with the case of the naturalist account of morality, emotion and sentiment – religious feeling – will be paramount. Wilson claims that such emotions 'clearly have a neurobiological source' (Wilson 1998: p. 288), and he notes the existence of a brain disorder in which the main symptom is the attribution of religious significance to everything within the sufferer's field of experience. Hence, he suggests, the theory that evolutionary processes have created a human brain in which there is an epigenetic tendency to religious belief is intelligible and consistent with the biological evidence, although that again shows nothing about the truth or falsity of the content of those beliefs.

He then suggests that a great deal of the characteristic behaviour associated with religion, such as those elements of propitiation and sacrifice to appease the gods involved in religious ceremonies, roughly fits the accounts of the dominance hierarchies so characteristic among mammalian societies. Modern human beings have clearly not broken free of this mammalian pattern and, particularly where religion is concerned, are readily attracted to follow charismatic individuals, especially males who bolster their natural charisma with claims of direct contact with the supremely dominant god or gods (Wilson 1998: p. 289).

The human tendency to create rich and aesthetically-powerful forms of culture is readily assimilated into the service of these powerful tendencies, as is the development among some individuals of techniques for attaining what is thought to be mystical contact with the gods and their merging into total communion with them. These are experiences that Wilson says he has undergone to some degree himself, during the period of his life when he was a 'reborn evangelical'. He is at pains to emphasise that even if they can be given a fundamentally evolutionary and genetic explanation, they should not be regarded as unimportant. They have been a key human preoccupation for thousands of years, and remain a compelling aim for many religious believers. The account given by St Teresa of Avila in the sixteenth century of her attempts to reach such a mystical union with God is cited by Wilson as an example of the power of this conception and of the precise form of consolation that it can provide for the individual human mind (Wilson 1998: p. 292).

All of these factors add up to a complex of enormous power. They account for the continuing hold which religion has on the human mind even in the face of the developing naturalist accounts of the origins of religious belief. He is fully aware of the phenomenological power of transcendentalism, in both morality and religion: 'Transcendentalism, especially when reinforced by religious faith, is psychically full and rich; it feels somehow *right*. In comparison empiricism seems sterile and inadequate' (Wilson 1998: p. 291–2).

The implication of this position, of course, is that a non-religious point of view is likely only ever to be attractive to a relatively few human beings. The intellectual arguments for non-religion, for a purely naturalistic view of the world, are clearly thought by Wilson to be compelling. However, it is important to repeat the point, fully recognised by Wilson, that although naturalism explains the origins of religious belief and its ubiquity and power, that in itself does not refute any of it. We might, on the face of it, have evolved to have an in-built tendency to believe something true with respect to the origins and purpose of the universe, although the sheer variety of religious beliefs suggests that at most we can only have evolved to have a tendency to believe some transcendental account or other.

However, Wilson takes the view that the empirical account recently developed by natural science, specifically by biology, of the origins of life is in fact incompatible with any religious belief. Some religious thinkers, aware of this fact in some degree, have sought to minimise the clash by rendering the idea of God more and more abstract and elusive. But, Wilson argues, the picture of the universe which science now presents, is one in which the role of a God is harder and harder to locate (Wilson 1998: p. 292).

The upshot of this discussion is, then, that Darwinian naturalism can offer plausible, though still somewhat crude (as Wilson accepts) explanatory accounts of the origins, ubiquity, power, and entrenched nature of religious belief among human beings. The content of the scientific, and specifically Darwinian, account of the origins of life then provides the intellectual case against the content of religious beliefs, and transcendentalism in general. Hence, Wilson in effect endorses the concerns of those religious believers, particularly in the USA, who see Darwinism as a threat to their beliefs, and does not accept the view of other religious believers, particularly in the mainstream Christian churches, that Darwin's theory of evolution is fully compatible with at least Christian religious belief.

Some of the opponents of Darwinism in the USA, recognising the logic of this argument, have sought to show that Darwinism must itself be a religion, for it clearly has direct negative consequences of the kind just noted for existing religious beliefs. This assumes that only a religious belief can refute another religious belief. The case for non-religion must then be itself a religious claim. But there is no good reason to accept this argument. The incompatibility between religion and Darwinism arises partly because some religious beliefs make factual claims about the origins and nature of the world which, if Darwinism is correct, are clearly false. But more importantly, it arises because at least some forms of religion suppose that only a transcendental form of explanation will be up to the task of causally explaining the universe, or aspects of the universe, satisfactorily. It is precisely this claim, at least as applied to the phenomenon of life, which is refuted by Darwinism. This rejection of transcendentalism is not itself a form of transcendentalism, and depends upon the extent to which explanations of the existence and character of natural phenomena can refer exclusively to empirical phenomena.

This is a matter that can properly be regarded, at least in principle, as still open to question. It is part of the case made by proponents of intelligent design theory that some observable phenomena cannot be causally explained on the basis of Darwinian presuppositions and theories. As we noted in the introduction to this book those examples do not seem to work. However, the important point for the issue of whether or not Darwinism refutes religion is whether the alternative offered to the Darwinian account is an alternative empirical account or a transcendental one. Only if the latter is offered does the adequacy of Darwinian naturalism fall under question, and only then is there space created for a religious explanation of the phenomena in question. This, of course, would be tantamount to showing that science construed as a purely empirical activity is inadequate even within the realm of causal explanation.

But if no scientific explanation is driven into the transcendental realm, and thus beyond science, then causal explanation is securely in the grip of science, and the intellectual route to the transcendental cannot follow that causal path. This leaves the spiritual path, concerning the meaning and purpose of reality. Many religious believers would now say that they are happy to leave causality to science, and give the meaning issue exclusively to religion. There would then be no possibility of a clash between religious and scientific truth, for each would be talking about different things. This, strategy, however, does not survive close scrutiny. Meaning and causality cannot be so easily divorced. For the universe to have a God-given meaning it is a necessary condition that God should have causally-produced it, and have causally-produced its key features. Otherwise there is not obvious meaning to the phrase 'God-given'. If natural science can produce causal explanations of the existence and character of the universe which do not involve a transcendental cause, then there is no intellectual basis for the postulation of a god, and so no intellectual basis for attributing a God-given meaning to the universe. If people still postulated such a being it could only be on non-intellectual grounds. Many would still be happy with this, and we have seen the powerful motives operating within the human mind, if Wilson and other Darwinians are correct, which would lead many still to operate with the religious worldview. But the intellectual pretension characteristic of religious belief would have to be given up.

However, many religious believers will be inclined to make the claim that Darwinism cannot get rid of religion as an intellectual construct that easily, and that Darwinism itself should now be seen as a religion. Let us consider that charge next.

Darwinism as a Religion?

Let us return to the issue with which this chapter began. Can Darwinism offer an explanation of pain and suffering which competes with theodicies without becoming a religion in its turn? We noted above that it makes sense from the Darwinian perspective for human beings to be glad that they have been brought into existence by blind natural forces, but that gratitude towards such forces is senseless, precisely because they are blind and operate without intention. It makes no sense, then, to offer prayer to such forces, or to seek to turn aside their 'wrath' by propitiatory acts. Their destructive outcomes are not produced by wrath, or any other morally assessable or affectable factors.

The natural world, the workings of which are so carefully explored and explained by Darwinian-directed biological and other scientific theories, is not, then, an appropriate object of worship, from the Darwinian point of view. Nor is the humbling of the human self before its creator, central to the act of worship in most religions, appropriate in this context. Neither pride nor humility are appropriate emotions to feel in contemplation of the natural world from the Darwinian point of view.

There are, of course, many important emotions which are appropriate in this domain. It is appropriate to feel fascination and awe in the contemplation of the natural world. The more we understand about its workings the more we find our imaginations and understandings being stretched and our presuppositions being undermined. It is appropriate, then, to feel humility in one way in connection with the natural world – humility with regard to our own knowledge and understanding of it.

The feeling of respect for the natural world is also an entirely appropriate emotion. Respect is always appropriate when we are interacting in important ways with entities whose character is in many ways unknown to us. The opposite of this kind of respect is the attitude of taking something for granted, assuming its character is fully known and accounted for, with no further elements within its nature to unearth. The immense, and probably intensifying, complexity of the

natural world means that we have to struggle to understand it. It confronts us as
a mysterious presence, which is not to say that it is a personal presence. We are
in many ways probably only beginning our voyage of discovery with respect to
the workings of the natural world, at least if we understand that voyage in terms
of theoretical understanding. In other ways, however, the natural world cannot
but be completely familiar to us.

It is, after all, the matrix of our own existence, and, as Midgley has often
been at pains to emphasise, in many significant ways we are completely at home
in it (Midgley 2002: pp. 95–8). Its character is a condition of our own self-
understanding. We are, thus, not lodged in a completely alien universe, possessing
the dark grandeur of the existentialist self, relying only on our own free will to
conjure fragments of meaning out of experience. The interaction we have with
the natural world is complex and enigmatic, but contains a large number of the
elements necessary for any human being to reach a sense of meaningfulness. The
world provides us with meaning every bit as much as it provides us with nutriment,
a fact celebrated by, among others, the writers of the Romantic movement. This
is not, of course, to deny that such meaning may on occasion run out for any
given individual, as may food and drink. We find the resources we need within the
natural world for our existence at all levels of our being, but nothing is guaranteed.
However, no other, more certain, source is available.

It may be appropriate at this juncture to mention the idea of faith. Midgley
has suggested that faith is 'the sense of having one's place within a whole greater
than oneself, one whose larger aims so enclose one's own and give them point
that sacrifice for it may be entirely proper' (Midgley 2002: p. 16). She remarks that
this kind of faith is plainly widespread and very important in our lives, and that
while not itself religious faith, for the faith in question may be in purely human
creations such as a political movement, it may form the seedbed of such faith
(Midgley 2002: p. 17). If these claims are sound, what can immediately be seen
is that Darwinism cannot properly form the basis of such a faith. Darwinism
certainly does show how human life can be seen as 'enclosed within a whole
greater than oneself'. However, this whole is not obviously capable of giving
point to one's own aims. For the whole in question is expressly said to be devoid
of an overall point.

Hence, if one can truly be said to have even the non-religious form of faith
identified by Midgley in Darwinian processes it can only be because one has not
properly understood the nature of those processes. The account of evolution
furnished by Darwinism furnishes key materials for the discovery of meaning,
and the experience of wholes which may provide the non-religious forms of faith
just mentioned, but they are not themselves the appropriate focus of such faith.
The materials in question concern the importance for human beings of interacting
with a complex world of entities with which they have co-evolved and with which
it is possible for them to experience some fellow-feeling.

We need, then, to emphasise these two central points about Darwinism: 1) that
it does not justify a worshipful attitude, akin to that of religion, to the natural

world, but does support respect, awe and fascination towards it; and 2) that it lends its support to the search for meaning within human life in ways that give the interaction between human beings and the natural world some prominence.

At this juncture it is appropriate to note the key distinction between spirituality and religion. Spirituality is the tendency of the human mind to seek meaning in the world and in its own existence. Religion is a form of spirituality in which the meaning desired is located in some supernatural order. Theism is the form of religion that characterises this supernatural locus of meaning in personal terms.

From what has so far been said about the non-religious nature of Darwinism we can conclude that Darwinism is nevertheless capable of sustaining to some degree the common spiritual urgings of human beings. It does so in virtue of the feature of scientific thinking which Midgley has highlighted, namely that science has to be conducted on the basis of some narrative vision, some overall, synoptic, view of what the world is like (Midgley 2002: p. 4). Human beings can find within such narratives some of the ingredients needed to furnish themselves with a sense of the meaningfulness of their existence.

However, in the case of Darwinism, the connection between overall vision and narrative on the one hand and meaning on the other is indirect. For the overall vision is not a vision of overall meaningfulness. It is a vision of how meaning-seekers came into existence, and of where it is appropriate for them to look for what must be always partial, shifting and contingent bases of meaning. It thus a vision that directs meaning-seekers towards rewarding sources of meaning, without directly providing such meaning itself. Hence, Darwinism has an important part to play in sustaining people's spirituality in a properly grounded understanding of the nature of existence, and does so in a non-religious way.

Unfortunately, Midgley herself sometimes fails to keep clear the distinction between religion and spirituality. Thus, she suggests that sometimes scientific doctrines can serve some of the functions of religion, because they are thought to be 'aimed essentially at the spiritual nourishment and salvation of the human race' (Midgley 2002: p. 14). She wishes, rightly, to criticise the use of scientific doctrines for this religious purpose. However, in the passage just cited she seems to be supposing that spiritual nourishment can be found only in religious views, and thus that in providing spiritual nourishment scientific doctrines are being converted into religious ones.

She shows herself elsewhere to be well aware that this connection between religion and spirituality is not necessary, and that human beings can find meaning in their everyday lives and in narratives which do not invoke some supernatural order, personal or otherwise. Darwinism is resolutely non-supernatural in its account of the origins and development of life on this planet. It therefore cannot really qualify as a candidate for being a religious viewpoint.

However, some, including Midgley, have seen lurking within the minds of some influential proponents of Darwinism tendencies which, they claim, formulate Darwinism as at least a quasi-religion. Let us consider some of these charges.

Midgley has long been a formidable opponent of the dangers of what she sees as one-sided thinking, especially when this concerns the nature of human beings. We have already encountered, and offered some critical reflections upon, her arguments in this vein in the earlier discussion of humanism and environmentalism. Although generally supportive of the biological approach to understanding human beings, to Darwin's evolutionary theory, and even to many of the hypotheses of sociobiologists, she has for some time been a stern critic of an overly-simple view of what these, in her view, indispensable modes of thought are.

Some of these oversimplifications have, on her account of the matter, been perpetrated by otherwise sophisticated proponents of the Darwinian worldview. She castigates sociobiologists such as Wilson, Dawkins and others for offering what she believes to be one-sided and thus highly misleading accounts of what the discoveries in genetics that are the heart of the modern Darwinian synthesis imply about human life. In particular, she is scathing about any attempt to support a debunking or reductionist account of human moral thinking on any basis solely to do with the gene's-eye viewpoint. She takes on some older targets too, such as the Social Darwinism of Spencer and others whose understanding of Darwinism was, to say the least, highly faulty.

We need not follow her argument through all its complexity on such matters. These points have all been addressed in earlier chapters. The issue of interest here is the diagnosis she offers to the effect that some of these forms of oversimplification, as applied to Darwinism, and perhaps too to other forms of contemporary scientific thinking, bring about what is in effect the transformation of such theories into something like religions. They are non-theistic religions, to be sure, but then, as we have just noted, religions do not need deities.

The two oversimplifying tendencies within Darwinism which she detects as producing this effect of creating a quasi-religion out of the theory rest on two intelligible and justifiable reactions to the account of nature furnished by Darwinism, attitudes which Darwin himself always struggled to hold in balance. They are attitudes that we have already noted in this chapter – 'on the one hand, optimistic, joyful wonder at the profusion of nature, and on the other, pessimistic, sombre alarm at its wasteful cruelty' (Midgley 2002: p. 5). From either of these two intelligible starting points we may be led to formulate a one-sided view of the matter which involves the dramatisation of one of the tendencies as the sole or fundamental truth about the nature of the world, with the other being lost sight of.

The dramatised view of the world that is spun out of the first strand is wildly and wilfully optimistic, taking our success in achieving some understanding of how the natural world has come about by Darwinian processes to prophesy that we human beings will shortly be in a position to take control of those processes, particularly at the genetic level. We will then be able to deal with those personal and social problems that derive from the unplanned nature of the natural world by the application of biotechnology, perfecting ourselves and our societies. The dramatised view that comes out of the second strand is equally wildly and wilfully pessimistic, positing an irredeemably hostile universe and/or a human nature that is unavoidably selfish. It is in this framework that the existentialist picture of the individual human self marooned in a sea of meaninglessness finds its natural home, and posits the gloomy magnificence of the absolutely free choice which is supposed to be our sole recourse.

She is, of course, entirely correct in what she says in the way of criticism of these tendencies, and by implication of any other such oversimplifications of the subject-matter in question. To take the pessimistic strain first, as we have already noted, nothing in the Darwinian account of evolution requires us to see the world as hostile in any absolute sense. We have evolved in this universe. Our characteristics as an animal species have arisen by the processes we can see at work all around us. Although the Darwinian account sees no purpose in our being here, and no guarantee that our species will remain in existence, still it also explains why the world presents us with opportunities and satisfactions, not just threats and pains. The Darwinian account also reveals the bases of our moral selves in our evolution as a certain kind of social animal, an animal that has it in common with other such animals that we have a tendency to behave altruistically towards at least some of our fellow species-members at least some of the time.

The optimistic side simply fails to see that scientific knowledge on its own however sophisticated and well-grounded it may be taken to be, cannot lead us to a perfect future in any humanly-meaningful sense. What moral thinking concerns is the complex balancing of claim and counter-claim in the light of a whole complex and shifting web of considerations – loves, loyalties, (justifiable) hatreds, hopes and fears and so on – which cannot be reduced to any algorithm. The Darwinian account of the evolution of human nature does not require us to abandon such complexity, or pretend that it can somehow be abolished. Elements of it are present at all levels of life, although it reaches its most complex phase (at least to date, on this planet) with human moral agents.

We should take note that Wilson, at any rate, shares her scepticism about the wisdom of seeking to use our increasing understanding of genetics to try to fashion a new, more perfect, form of human being. He does so by putting forward a defence of conservatism with respect to changes in the human species, where conservatism is taken to mean '… the ethic that cherishes and sustains the resources and proven best institutions of a community' (Wilson 1998: p. 309). He predicts, approvingly, that if human beings are given the choice about how to deploy their genetic knowledge, they will happily, and justifiably, use it to cure clearly-identifiable illnesses and undeniable handicaps which have a genetic basis, but will resist trying to produce more perfect people by altering humanity's heredity. For our heredity is the physical basis of what it is to be human, and being human is what is precious to us, imperfections and all. This clearly chimes in with Midgley's general view that we do have an evolved nature, and that the demands of morality, and the roots of human happiness, both involve understanding and respecting it. In addition, Wilson is clearly sceptical about the ability of human beings to deploy such genetic engineering successfully – 'Neutralize the elements of human nature in favor of pure rationality, and the result will be badly constructed, protein-based computers' (Wilson 1998: p. 309).

Darwinism is an elegant solution to the question of how life evolved on this planet. But it is not a simple account when applied to any given example of such life-forms. The elegance of the general solution, coupled with a tendency to focus on only one or other undeniably present feature of life, is what may lead to the erection of a quasi-religious version of Darwinism, in which ultimate truth has finally been vouchsafed to us and in which we may finally put all our trust.

But the Darwinian worldview need not acquire this quasi-religious character. As a scientific view, it should always be subject to the postulate of fallibility. It may in fact turn out to be false. We may find incontrovertible evidence that species did not evolve by natural selection, or evolved by some alternative mechanism, or did not, after all, evolve at all. We would then have no alternative but to abandon the theory. This may becoming a remoter possibility as the theory of evolution by natural selection seems to gain, rather than lose, evidential support. But it is a possibility that can never be eliminated.

This point alone should help prevent the erection of Darwinism into a quasi-religion, and when taken in conjunction with the points already noted about the importance of avoiding one-sided, simplified, accounts of the phenomena crucial for the generation of religious belief – especially human moral thinking – ought to be sufficient to guard against the whole phenomenon.

It is important, too, to guard against a different tendency, this time one which may be teased out of the position adopted by Midgley, although she does not do this herself. This is to argue that because Darwinism is subject to fallibilism, because reality is complex, and because Darwin's theory of evolution by natural selection cannot explain everything in human life, or even life in general, then it does not have to be taken seriously as an account of the fundamental features of human life. This book has been an attempt to show the way in which, in spite of Darwinism's character as a scientific theory, subject to change and development, it promises the possibility of an important contribution to human self-understanding and to the understanding of the natural world in general which cannot easily be brushed aside. It is perhaps a temporary platform on which to stand, but that is the only kind of platform we have available, and this one may well be more durable than any others we have so far found.

Chapter 10

Conclusion

Consider the following parable. A certain elderly gentleman, called Darius Wynn, was one day walking along the sea cliffs near his village, as was his regular practice. A lifelong lover of nature, he was following a little-used path which led him to an area of the cliffs where he could observe undisturbed the birds, insects and flowers which had always fascinated him. It was a beautiful spring day, and the air was filled with the humming of bees and the singing of the lark. The edge of the cliffs was bedecked with a profusion of wild flowers and in the sea, above which soared and glided various species of seabirds, he just caught sight of a seal's head before it dipped below the waves.

He paused at the very edge of the cliff, raising his binoculars to get a closer view of a butterfly that had landed on a plant someway down. Suddenly, the ground on which he was standing gave way, and Darius, losing his footing, tumbled down the steeply sloping ground, hitting his head hard on a rocky outcrop that knocked him out. When he regained consciousness he found himself lying on his back, his descent arrested by a sturdy bush. To his horror, he found himself unable to move any part of his body below the neck. He could turn his head, with effort, from side to side, and became aware that he was lying some way below the edge of the cliff, well out of sight of anyone who might be walking along the path.

He lay there in this paralysed condition, able to call out for help, but aware of the effort this caused him, and aware, too, that his elderly voice was unlikely to carry very far. The lovely spring morning carried on around him. The sun still shone brightly, and he could hear the distant skylark delivering its aerial rhapsody. The butterfly he had wanted to see more closely fluttered nearby, alighting a few yards away, so that he could easily ascertain its species. As he turned his head, he found near his face clumps of the spring flowers, occasionally fluttering in the gentle breeze, which wafted their faint perfume into the air. From time to time a gull glided into his line of sight against the blue sky, sometimes calling out before moving on in pursuit of its own purposes. The buzzing of a bee grew louder in his left ear. It alighted on a flower to collect its load of nectar and unwittingly deliver a dusting of pollen, before resuming its humming trajectory across the face of the cliff, the sound rapidly diminishing with distance. Flies landed on his face, increasingly covered with sweat from the heat of the day and his own rising anxiety.

In this condition the beautiful world of nature, his lifelong love, suddenly took on a wholly discomfiting aspect. All the living objects of his devotion, concern and compassion were there, just as they had been a few minutes ago, before his fall. But now they were revealed, with an almost hideous force, to have absolutely no interest in him, except, perhaps, as a perch or source of moisture. This lack of interest, of course, was not actual hostility – more sublime indifference. Darius simply did not figure in the calculations of the bird or insect brain. The richly interconnected world of living things which he had always cherished and striven to understand, and which he believed himself, and all human beings, to be part of, simply had no concern for, or understanding of, him. He could die at that very spot, his body could rot away and be devoured by bacteria, insects, birds and mammals, but his existence as a person, an intelligence, a moral agent, a being of rich and complicated emotions including love and hate, fear and hope, would remain completely invisible to that company of undertakers.

Of course, Darius had always understood that while human beings can, and do, care – often passionately – about their living planet, the planet is incapable of caring about them. But now it had been brought home to him with inescapable force that the solitary individual, like Darius paralysed on the face of the cliff, is highly vulnerable at all times. Nature may furnish resources and opportunities, but it also offers threats and harms. Human beings, it was now apparent to him with awful clarity, can rely only on each other for help in extreme predicaments, such as the one in which he now found himself. What had hitherto been a largely intellectual proposition was now delivered with the force of a punch to the solar plexus.

Darius had never been a misanthrope or in any way antisocial. But he had rather taken the existence of human society and human solidarity for granted, living as he did in a civilised and affluent land, in which hospitals, doctors, firefighters and police were always on hand. But now all those things seemed as far away as Alpha Centauri.

After an hour or so of isolation and dread, with mounting thirst and feelings of faintness that he struggled to fight off, Darius suddenly heard the unmistakable, if still distant, sounds of a human voice. It was singing in an uninhibited way, which, as the voice grew louder, assumed a clearly inebriated character. Darius realised that he was listening to the raucous anthems of the village drunk, Hugh Mann. Hugh had never held down a job for any length of time, had fathered several illegitimate children and, when sober, was believed to scratch together a living out of sponging off of relatives, petty thieving and welfare fraud.

As the voice drew level with Darius it stopped, and then resumed as a low, tuneless drone. Hugh sounded as if he were preoccupied with something. Then silence for a few seconds, before Darius saw, above the edge of the cliff, a flash in the sky which resolved itself into a glass object tumbling through the air and bouncing down the steep incline just below his feet. Clearly, Hugh had drained his bottle and thrown it away over the cliff in one of those careless acts of petty vandalism which typified (so the respectable villagers like Darius had always said) low life such as Hugh: no respect for the natural world or for other people's appreciation of its beauty.

With a supreme effort Darius called out for help. After a moment a face appeared over the edge of the cliff, against the blue sky. Hugh's drink-sodden eyes focused on Darius after a few seconds, gazing blankly upon the prostrate form held in place by the bush below him, before comprehension gradually filtered into Hugh's brain. 'Are ye all right?' he called out, in his slurred voice. 'I can't move. I'm badly hurt', gasped Darius. 'Right ... Right', said Hugh, as his mind slowly clicked into gear. 'Don't ye worry ... I'll get help.' Darius saw the face disappear and a loud belch from some distance away told him that Hugh was on his way.

This parable shows, of course, that the one thing that human beings are always capable of offering each other, whatever their shortcomings in terms of moral ideals, is simple care and concern. This may not always be forthcoming. Hugh in the parable might have been Darius' worst enemy, or some form of psychopath, or just too drunk to be any use. But if care and concern for Darius in his plight was going to come from somewhere, it was going to be from some human being or other. True, occasionally there are reports that people's pet dogs (usually not their pet cats or pet tortoises!) have played a key role in their rescue from some predicament similar to that of Darius. However, this seems to be because the dog's reaction to its owner's plight stimulates unusual behaviour in the animal that attracts the attention of some human being to the situation.

Sometimes this is given a, usually implausible, anthropomorphic gloss, to the effect that the dog itself diagnoses the situation and, moved by care and concern for its owner, consciously seeks human aid, as in the 'Lassie' movies. It may not be impossible that at least some of these accounts are true, for our understanding of the minds of animals other than ourselves is still at a rudimentary stage. But even if one does accept at least some of these stories, human beings remain overwhelmingly the only reliable source of care and concern for other human beings.

This parable is not meant to deny that members of other species are capable of showing at least some forms of care and concern, or perhaps at least behaviour which involves taking care, with respect to members of their own kind. Parents and offspring in many complex animals show this kind of behaviour. Also, in our closest primate relatives there has been observed what it is not implausible to characterise as concern shown by one member of a group for another's injury or distress, even when they are not directly related. However, this is usually transient and not developed very far.

The key point remains that human beings alone are able to take a deep and lasting interest in the welfare of other species of a kind which cannot be reciprocated by those other species. Which is not to say that we always have such an interest, or an interest that lasts, and some people never seem to develop it. Many seem to have purely instrumental attitude to non-human life-forms in general. But it is a perennial feature of human beings, across time and social distance, for some form of such concern to develop.

However we explain this human capacity – and perhaps some form of unsustainable anthropomorphism is necessary for it to be generated and developed

– the asymmetry of concern between human and non-human highlighted by the parable may be used to drive home a point against the Darwinian worldview. For it may be argued that it is important not to draw the wrong conclusions from the fact that human beings share a common descent with all other living beings on the planet, and that we are thus literally cousins of the chimpanzee, lion, eagle, snake, shark, spider, earthworm, lily and bacterium. For, these and all other life-forms can care nothing for us and our well-being, even though many of us care greatly for them and theirs. It is important, then, it may be argued, not to be carried away by some sentimental attitude to the non-human world. The latter may be fascinating, but there is a fundamental gulf between the human and all other species. We are not really part of their world, whatever the biology seems to reveal about the connections between us. Moral agents can only really interrelate in any full-blooded sense with other moral agents. Beings that can be no more than moral patients, however complex and fascinating they may be to moral agents, are separated from the latter by a crucial moral and ultimately, therefore, ontological gulf. That is, the kind of being a moral agent is differs profoundly from the kind of being a moral patient is.

The Darwinian worldview, it may be argued, seeks to deny, or at least to obscure the existence of, this ontological gulf. It seeks to do this in two dimensions. Firstly, with respect to the explanatory project, it seeks to assimilate the understanding of human beings – their mental structures and consequent behaviour – to the understanding of non-human beings. This, it may be argued, is a tempting direction in which to head if you are mesmerised by the Darwinian account of the origin of species by natural selection. For this is likely to plant in your mind the conviction that the understanding of human beings just has to be developed along the same lines as the understanding of non-human species. The idea of any kind of a gulf will sit very badly with an evolutionary narrative which constantly emphasises continuities, connections, developments out of common ancestral forms and so forth.

From this standpoint, the whole idea of human sociobiology, in whatever form it is cast, is radically misconceived. The crucial fact about human beings is highlighted by the parable of Darius Wynn and Hugh Mann. These two interrelate in a unique, vitally important and defining way, whatever the psychological and cultural traits that distinguish them from each other, in terms of their values, beliefs, behavioural tendencies and the rest. They recognise each other, however dimly, as inhabitants of a common, moral, universe. They recognise, in so doing, that they have duties and responsibilities to each other. This recognition does not operate between members of other species, even when they do show the limited, instinct-based, capacity to take care of each other. Human beings can extend this recognition to non-moral agents within the human sphere, such as new-born babes and the severely demented, and beyond, to other species. But the bed-rock distinction is between human moral agents and all the rest.

From this fact all else follows. Human culture, acknowledged by the gene-culture coevolution theory to be of central importance in the development of human phenotypes, is quintessentially a moral phenomenon. The narratives central to cultural self-understanding are all tales of taboos broken, promises shattered, revenges and punishments exacted, rewards for virtue bestowed, justice withheld then restored and so forth. The narratives provide the contours of culturally shaped behaviour. The explanation of such behaviour must, therefore, refer continuously to a central phenomenon – morality – that is simply absent in the non-human case. Biology may have produced human moral agents, but the forms of explanation adequate to deal with their behaviour must definitively cut free from the biological pattern. Consilience is, therefore, a gravely mistaken concept.

Of course, moral behaviour of this sort can only exist if other elements within morally structured minds are also present. Self-consciousness; means-end rationality; the ability to manipulate the physical world to fit specific, consciously-devised, purposes – and thus a high degree of technological ability; language-use for storing and communication of knowledge, and so forth must all be present. Some of these may characterise, at least to some degree, non-human species. But until they facilitate the transition to the moral dimension the ontological gulf has not been crossed.

The second element in the Darwinian worldview that the parable may be taken to undermine is the claim that human moral agents have moral responsibilities towards, not merely with respect to, the non-human world. This is a view that sets out to obliterate the ontological gulf in another dimension, by seeking to draw non-human beings into the same moral universe as human moral agents.

Darius and Hugh are very different in terms of the status they hold within their society. Darius is respectable, compassionate, concerned about the welfare of others, not just of his fellow human beings, but also of the non-human world. Hugh is feckless, irresponsible, self-centred and parasitic on the efforts of those around him. Yet Hugh can recognise the claim on him that Darius has in his hour of need, and can exercise himself sufficiently to bring help where it is needed. Hugh may have serious defects as a human being, but he acts as a full member of the moral community. By contrast a seagull or a butterfly, however perfectly they may embody their species character, do not inhabit that moral community because they cannot. They are not at fault or in any way failures. They are just not that kind of thing – the ontological gulf again.

Hence, it may be claimed, an environmental ethic which takes its inspiration from the Darwinian themes of the interconnectedness of living forms and their common descent fails to respect this key difference. Granted, it may be a moral virtue of human beings that they take steps not to visit unnecessary suffering upon, or wantonly destroy, the non-human sentient creatures which they encounter, interact with, and often put to use. But that is not the same as granting such beings any form of moral status comparable with that of human beings who are moral agents. Any mode of thought which takes a different view from this is simply mistaken, and fails to note the ontological gulf operative here (see Lynch and Wells 1998 for an argument of this form).

In sum, then, the Darwinian worldview outlined in the course of this book is profoundly mistaken in both of the dimensions investigated. In both cases a key ontological dimension has been ignored or obscured.

The ontological distinction highlighted by the parable may, of course, be used as a basis upon which to develop further distinctions and ground further claims. For example, the unique status of moral agents may be developed along religious lines, so that the origins of this status are located outside the empirical world entirely. On this view, qua moral agents human beings are made by, in the image of, and answerable to, God (Rachels 1990: p. 4). But this is not a necessary development. It is possible to remain content simply with the claim that, however the status is question is to be accounted for, human beings are just not the same kind of being as other organisms. They cannot be re-categorised simply as a certain unique kind of organism. Their moral nature colours everything else about them. It is a nature best explored by the methods of the humanities and social sciences, focusing upon human moral phenomenology and its structuring through specific cultural forms, such as narrative, rule-making and norm-construction. Biological approaches will get us nowhere, because they are blind to the ontological gulf. In failing to respect this gulf they will undoubtedly lead to distorted understandings of what is characteristic of human beings, as the latter are wrongly assimilated to the non-human cases which they faintly resemble.

It will be readily apparent that the parable of Darius and Hugh has here served as a means to encapsulate and recapitulate the main anti-Darwinian arguments considered in the earlier chapters of this book. How ought a proponent of the Darwinian worldview to respond to them?

There is clearly nothing in the actual parable itself with which a Darwinian need take issue. As Darwin himself emphasised, human beings clearly are, contingently, unique in their capacity to show interest in and concern for other species from a moral point of view – although not just from that point of view (Darwin 1901: p. 193). The interest that Darius shows in the natural world is as much intellectual, aesthetic and probably in a basic sense spiritual, as it is moral.

Hugh and Darius are certainly also related to each other in a way in which neither relates to the rest of the living world. They are capable of mutual moral recognition.

Again, the rest of the natural world is certainly completely indifferent to us, insofar as it makes any sense to attribute any attitude to the non-human. Ironically, in the light of the parable, for many people one of the spiritually important facts about nature that forms an important basis of their interest in it is precisely that it is independent of, and uninterested in, human beings. This allows them, they believe, to put their own problems into proper perspective after they have sought out an encounter with the non-human natural world.

Certainly, of course, we are the objects of some kinds of interest for some kinds of organisms – as prey or hosts or symbionts, and so forth. But this is not to attribute any form of concern and care for us for our own sakes. The Darwinian account of the mechanisms of natural selection does, of course, provide an explanation of this indifference. For it describes a blind, purposeless process in which all the life-forms which have emerged can be shown to have come about without any prior plan, and therefore not as the result of a planner's malevolence or benevolence. In such a world there is no prior reason to expect mutual concern amongst organisms.

If, nevertheless, such concern does start to manifest itself, then this calls for an explanation. Once again, the Darwinian account attempts to provide one. It can account for the emergence of altruism, firstly amongst members of the same species, and then between species. It also accounts for the emergence of creatures like Darius and Hugh, moral agents who develop a sense of mutual recognition and responsibility. It does this job of explanation partly by relating the species-characteristics of beings such as Darius and Hugh to those of other similar types of organism. Social species that develop the kinds of intelligence characteristic of human beings, as we saw Darwin himself to argue, will strongly tend to develop the forms of moral thought and behaviour characteristic of human beings.

Hence, the Darwinian perspective starts to reveal the explanation for features of the parable which anti-Darwinian accounts have to take as givens – as reflecting merely brute facts of ontology. It can also start to develop explanations for another key feature of the parable, which is the difference of moral status and outlook of Darius and Hugh. The reasons why human beings develop different personalities, why there should be differences between particular phenotypes at all, what differences there are likely to be, and how they are likely to be evaluated within human societies are all subjects for investigation, and explanation, from the Darwinian perspective.

In the course of these investigations the point upon which the anti-Darwinians are inclined to lay such weight – the ontological gulf – has to be rejected just because it blocks explanation. Positing an ontological gulf converts a phenomenon that might otherwise receive an explanation into a brute fact of existence. There are undoubtedly going to be such brute givens, but they should be reserved for the genuinely ultimate levels of explanation, and what such levels are is far from clear yet. The natural sciences are still investigating how to characterise the ultimate modes of existence, and the old, time-worn philosophical/scientific categories of matter and mind, atoms and the void, are plainly no longer of any use.

The gulf itself does genuinely begin to fade away upon closer inspection, from the Darwinian point of view. To begin with, the moral recognition which human beings accord each other is not an all-or-nothing affair. It admits of nuances and degrees. The help which human beings are prepared to accord each other has a definite shape, moving from the duties to kin and close relatives, to fellow-society members, to aliens. Different moral theorists have characterised this shaping in various ways, although all have noticed it – some to endorse it; some, such as cosmopolitans of various forms, to criticise it.

The Darwinian approach expects such variations, and has the resources, once again, to offer at least the beginnings of an explanation of them in terms of the development of human moral sentiments out of the prior phenomena of kin-selection and reciprocal altruism. The phenomenology of human moral experience is thus not taken as a given, but is categorised and investigated by means of concepts which only emerge when the idea of an ontological gulf is abandoned, and human beings are located firmly within an evolutionary framework in which their continuity with other life-forms is kept fully in the forefront of our awareness.

With respect to the second strand of criticism, directed at the attempt to encompass the non-human, at least in part, within the ambit of moral concern, Darwinians, as we have seen, are divided in their approach. Wilson's environmental ethic was, as we have seen, pretty relentlessly anthropocentic, even though from time to time he recognised the possibility of granting to the non-human a clearly moral status. In general, for many Darwinians the second strand of criticism by the anti-Darwinians will not be objectionable at all. They will happily accept that the line delineating the moral community should be drawn so as to encompass only one species – *Homo sapiens.*

But arguably the spirit of the Darwinian project should lead Darwinists to be sceptical of this. As we have seen, for the Darwinist the project of explaining morality among human beings cannot have truck with the idea of an ontological gulf between human beings and other organisms. The continuities upon which the Darwinian account relies must really be there. But, if there is no such gulf, then the claim that there is a hard and fast line separating the species which count morally (only one) and those which do not (the rest) ceases to have any ultimately compelling quality. It cannot be anchored in any objective, given, difference of ontological status. If so, then there is no reason to expect that human moral thought will track such a status. The fact that Darius has moral concern for the non-human world then ceases to be a puzzle. Darius has noticed the similarities between the human species and the non-human, and has found this to be a reason for attributing at least some forms of moral status to them directly (an implication of Darwinism that has been explored by James Rachels (1990)). In this respect Darius is far from alone, for most cultures have so drawn the moral line that at least some non-humans are included, albeit for a wide variety of reasons (Baxter 2004: p. 2).

The Darwinian account, then, makes it intelligible why human beings who are not obviously deranged have found compelling reason to extend moral concern to the non-human. There nevertheless are key differences between human beings and non-humans that it is reasonable for moral thinking to take account of. The understanding of such differences is itself a matter for further investigation, interpretation and disagreement between reasonable people, which is why there is need for the field of environmental ethics in the first place.

Having surveyed the general case for and against the Darwinian worldview in the light of the parable of Darius and Hugh, let us finally review the contributions to these debates of Edward O. Wilson.

Wilson must take a great deal of credit for raising to prominence the whole idea of the application of biological thinking, having Darwinian evolution at its core, to human beings. In this respect he boldly pushed on where others were hesitant to tread, playing a notable part in the development of the sociobiological family of theories. In so doing he has repeatedly pressed for the re-examination of presuppositions about what human beings are, how they relate to the rest of life, and how best to attain an understanding of them.

In doing this he has inevitably found himself challenging the conceptions and arguments of many, often extremely influential, individuals and schools of thought from a wide variety of human cultures. These are the approaches to human life which emphasise differences between human beings and other life-forms. They encompass nearly all religious forms of belief, several ideologies, much philosophy and the dominant strains of thought in western academia centred on the social sciences and the humanities.

But the challenge to these powerful traditions mounted by Wilson needs to be understood carefully. There is much in these various schools of thought which is preserved in the Darwinian perspective, just because many of the often-noted differences between human beings and other living forms are undeniable. Human beings do possess enormously larger and more complex brains for their size than those of other animals. The possession of this brain makes possible self-consciousness, language, morality, elaborate culture, technology, the radical transformation of the human environment and, via various feedback processes, the possible transformation of the human species itself. In the term favoured by the Hegelian and Marxist tradition, human action is praxis, remaking ourselves to some degree as we make over the world.

Wilson has been at pains to preserve these forms of distinctiveness in the sociobiological approach to human beings, and to deny that we can understand human beings as if they were just another species of primate. The distinctive traits of human beings, above all the language-culture complex, mean that there are whole dimensions of human life, unique to our species, that require analysis and understanding before we can begin to grasp the human phenomenon. Hence his attempt to develop the version of human sociobiology which allows scope for the interaction of culture and genome and allows, too, for the possibility of rapid evolution in human brains. The phenomenon of praxis is thus not ignored, but taken into account in the development of an evolutionary account of the development of the human brain.

He has also remained sceptical of those who have tried to assimilate some primate behaviour too closely to what is, probably, related human behaviour. For example, he is clear that chimpanzees do not have a version of human language in their wild state, and that the signs they have been taught by their human mentors do not have any of the key features of human language (Wilson 1998: pp. 144–5). Tool use and the elements of cultural transmission can be found amongst them, but it is all very rudimentary by comparison with the human case.

But he has been more than willing to grapple as and when necessary with those forms of the uniqueness claim that do not seem to him to be justified, and which the Darwinian approach does seem to undercut. The guiding thought in his approach, as indeed it is for all thinkers attracted by naturalism, is that human distinctiveness is not the same as, and does not logically presuppose, human disconnectedness from the rest of life. The two chief forms of the disconnectedness claim – versions of the ontological gulf we have already encountered – are moral and religious. In much moral and religious thought we encounter the idea that human beings are forced to recognise the existence of some transcendent, vitally important, realm of reality with which they alone have an intrinsic connection. The facts of human moral and religious consciousness are cited as key evidence in these areas. The fact that many people claim to have revelations and direct experience, albeit in a non-sensory way, of such a level is taken as evidence for its existence.

The case for transcendentalism in morality also derives much of its power from what appear to be important claims about the experience of human moral thought. It is often said to have an objective, compelling character that can only properly be accounted for on the basis of its revelation of the transcendental origin of moral imperatives. Human beings experience themselves as possessing a form of autonomy with respect to their moral deliberations which cannot be accounted for except on the basis that, qua moral agents, they are not ultimately to be regarded as embedded in the physical world at all.

In these ways, the phenomenology of human experience is used as a final touchstone of the ultimate nature of reality. Wilson is respectful of these phenomenological facts. He often recounts to us his own versions of such experiences during periods of his life when he accepted religious belief and the associated transcendentalist moral positions. He takes them to be so powerful and recurring, so important for the effects they have on people's actions as individuals and groups, that they are unlikely ever to disappear completely. He is aware too that they have produced fortunate as well as unfortunate effects in human life and have undeniably been at the core of important cultural manifestations for a wide variety of societies.

However, their use to bolster human exceptionalism, in the version which posits an ontological gulf between human and non-human, has the effect of placing an unjustified limit upon the scope of the Darwinian perspective. If we take the phenomenology at its own valuation then we are left with only mysteries where we might, if we tried, find explanations. We will, of course, probably always have unanswered questions, for every satisfactory answer we give raises new questions. But intellectual progress has only been made when some hardy individuals have sought answers to what others have been only too happy to label 'insoluble mysteries'.

The existence and frequent character of moral and religious thought can be explained in Darwinian terms, as we have seen. The explanation does indeed involve the rejection of the transcendentalist viewpoint and its replacement with a naturalist one. As we saw, the latter can explain why the transcendentalist phenomenology is experienced, and show how it might be replaced by the naturalist one, albeit with a certain amount of difficulty. This is because the human tendency to posit objectivity in these two areas usually, from the Darwinian point of view, bestows an evolutionary advantage upon those who possess it. Hence, the Darwinian perspective finds itself in the awkward position of seeking to reject theoretically a mode of thought the existence and power of which it can happily explain in its own terms. This is not a formal self-contradiction, but it does show clearly how Darwinism can reveal important difficulties in the implementation of its own project.

However, these Darwinian modes of explanation of religious and transcendental moral thought do not in themselves, as Wilson is fully aware, show that any particular example of such thought is untrue. Although the latter modes of thought are, logically-speaking, contraries, and thus cannot all be true (for they offer rival accounts of religion and morality), and although they all may also turn out to be false, still reason has to be given for believing that any one of them is false. The Darwinian perspective in itself does not do that.

However, it does seriously undermine the two traditional bases for giving any credence to their truth – the reason-based and the phenomenological foundations. The arguments based on design, which made most of the great Enlightenment thinkers accept Deism, have been seriously undermined by Darwinism, in spite of the best efforts of the proponents of intelligent design to argue the contrary. The phenomenological arguments can no longer rely upon taking the nature of the experiences in question at their face value, and the Darwinian perspective produces plausible reasons for why they should possess the character upon which so much argumentative weight is placed.

The alternative metaphysical view which is opened up by the writings of Darwinists such as Wilson has few of the traditional comforts offered by their transcendentalist opponents, which is not the same as saying that it is devoid of any forms of comfort at all. There is no good reason, from this point of view, for believing that human life, or any life, or indeed the universe as a whole, has an overall point of purpose. Point or purpose is only found within the universe amongst living creatures which have evolved to have such things, for those which have them survive better than those which do not. All their particular points or purposes have to fit in with the overall goal of survival, for if they undermine that goal the creatures do not survive, and their species makes a quick exit. Even the religious sense, at least in most cases, is strongly oriented towards the survival of death by human beings, in the face of overwhelming physical evidence that they do not in fact do so.

More specifically, then, the universe has no overall *moral* point or purpose. Morality, too, is only found within the universe amongst living creatures that have evolved to manifest it. As before, the point of morality has to have a strong orientation towards survival and the avoidance of harms which threaten that survival, amongst creatures with a strongly social orientation. On this view, as we saw earlier, and as Wilson has clearly argued, human morality is a contingent phenomenon and its character is heavily conditioned by the role it evolved to play within a certain group of primates. This implies the possibility of the evolution of a different behavioural tendency amongst future descendants of the human species that would have a similar role to morality, but not with its current general form. It is possible, too, that creatures elsewhere in the universe just as social as we are, but with a different biology, might well develop a distinctively different form of morality, experienced by them as having the objective force and transcendental quality so often attributed by humans to their forms of morality.

This is a universe, then, in which human beings have happened to arrive by the evolutionary route. That route guarantees that they will find themselves at home within that universe in many important respects. Fashioned out of earlier forms of organism by natural selection, they can only have evolved and survived by responding adequately, and better than their competitors, to the environment that has done the selecting. They must, too, have evolved a highly developed social sense – the sense of belonging in intimate relationships to others of their kind – that is the sine qua non of their survival and reproduction. In other words, human beings have evolved to experience, within the world in which they reach maturity, a sense, strongly mediated by social relationships, of being 'at home'. This is at least a partial consolation of the evolutionary account.

But, since this has been brought about by the blind workings of natural selection, the sense of home is never total. The forming of human mentality by natural selection comes about through what happens to enhance inclusive fitness, and this necessarily involves the environment's production of threats to life and well-being. If it did not do that, there would be no opportunity for forming or shaping of distinctive life-forms at all. But threats do not have to be perfectly dealt with. It suffices, from the point of view of natural selection, for a species to cope well enough for the main threats to its survival to produce viable, fertile offspring for the next generation. We have not needed to evolve eternal youth or immortality, and we have not. The world, therefore, remains a source of threats, and this fact is inherent in the vary nature of the universe within which we have evolved.

Self-conscious, intelligent, social beings are bound to recognise these facts inherent in the very idea of evolution by natural selection, even if it has taken a very long time for the human species to formulate the idea of such a process and hold it before its gaze. Casting round for an explanation of the fact that we are not fully at home in this universe we have formulated appealing stories concerning a future realm where we will be completely at home. Asking why we are not already in it, it has seemed plausible to many to hypothesise that this world represents a punishment or a trial, with the eventual arrival in, or return to, our real home a reward for our efforts. Casting round for the reasons there could possibly be for such a dire punishment, the other salient fact about us, that we can act well or ill, that we do as a matter of fact visit punishments and provide rewards for each other in our moral lives, lies readily to hand. The punishment is for a moral failure, or the trial is a test of moral virtue.

For our morality, too, viewed as an evolutionary phenomenon, bears the hallmarks of its origins. A phenomenon that arises by evolutionary processes has to be satisfactory from the point of view of enhancing inclusive fitness in the context of its development, but no more than that. Our moral sense, according to the account of sociobiologists such as Wilson, arose out of precursors that we can still see operating today within our primate cousins. Our social tendencies of care and concern for others of our kind does not have to be perfect to enable us to get an evolutionary advantage from it. It can, and plainly does, coexist with tendencies with which it is at war to greater or lesser degrees within human individuals, and to different degrees at different times of their lives.

Awareness of this leads to an awareness of internal division – of a self-centred and anti-social, and an altruistic and caring, dimension to our behavioural repertoire. We are not perfectly moral, though our moral phenomenology leads most of us at least sometimes to experience moral requirements as possessing a compelling quality. These two features of our phenomenology are encapsulated in the idea of original sin, or a prior falling away from an ideal to which we ought to have adhered. The punishment is thus readily attributed to a moral failing of human ancestors, although the precise nature of that failing admits of different interpretations, given the complex nature of morality. It may be a failure of action, feeling, obedience or understanding, or some or all of these together.

The Darwinian worldview has no need of this kind of account of the curiously equivocal relationship we have with our universe. Our feeling of being at home, of finding the world around us a source of delight and happiness, and yet also of its being a source of pain and upset are just what the Darwinian theory might lead us to expect. This at least offers the consolation of knowing that we are not in this situation because of anything dreadful 'we' have done. It can lift from us any burden of guilt we may be experiencing. It also has the consolation of letting us see that the loss of belief in these traditional explanations does not itself threaten the viability of human moral thought. It is not true that if god is dead, then everything is permitted. We are not, and cannot be, perfect altruists, but, if the Darwinian account is correct, we are not, and cannot be (at least most of us, most of the time) monsters of egoism either.

What we do have is the unavoidable struggle to live with ourselves, given the complex, morally oriented, nature which evolution has bequeathed to us. As we saw in the course of our earlier discussion of naturalism in morality in the company of Wilson, the key aim of people who take their understanding of morality from the Darwinian account will be the attaching of crucial importance to integrity and wholeness of being. Given what we are, given that we cannot avoid attaching importance to moral behaviour, on our own part and on the part of other people around us, we cannot but seek to get our individual psyches into such a condition that that behaviour becomes possible and normal for us.

We thus have to find some way of integrating our consciousness into a morally-defined shape, in the clear recognition that we also have evolved tendencies which pull us in other directions, but that these, though they can be recognised, cannot be allowed to dominate, even if they can be allowed to find outlets which are minimally disruptive of the integrity we are trying to develop. It will be a great help, from this Darwinian perspective on morality, to learn to accept these different aspects of our personalities rather than to become crushed by guilt with respect to them, or resentful of the need to deal with them. Arguably this picture of the moral life as an exercise in integration and the development of integrity in the face of the kind of beings we have evolved to be is a consolatory one. It presents important challenges, of course, but it does not waste time in vehement condemnations of the very fact of what we are – though vehement condemnation of our failure to integrate ourselves into a morally controlled form, or even to make the effort to do so, can still be found an important role. Morality understood from the Darwinian perspective need not be an anaemic affair. Understanding does not imply loss of feeling.

There remains, of course, the larger fact of human mortality, and of the possible consolations which Darwinism may be able to offer us with respect to this fact in the light of its picture of a universe in which we only arrived here by chance, we have no obviously privileged position, and where we certainly must come to an end as individuals, as quite possibly must our species as a whole. The consolations in this area are probably not great, or at last may not be very great for most people. If we had evolved to be unperturbed at the prospect of death, to accept it as naturally as we accept the sunset, then there would not be much of a problem. But we only survive because we flee the threat of death, we fear it in the same visceral way that many people fear spiders and snakes. Indeed, if Wilson's speculations on biophilia are correct, then fear of death is what underlies biophilia – love of life and fear of death seem as integrated as yin and yang.

Of course, as is always the case with human traits, not everyone fears death to the same degree, or to the same degree at every point of their lives. Those who have had a long and happy life, in which they have largely been able to develop what they have it in themselves to become, seem often to accept death with relative equanimity. For such people the point of further existence starts to become obscure if you can only make enjoyable use of it by first becoming a wholly different person. Nevertheless, there are good evolutionary reason why we should fear death, just as there are no good evolutionary reasons for us to be immortal. This juxtaposition of facts is what causes the problem.

Darwinism, however, may have a valuable role to play in focusing the mind upon this life and upon what forms of fulfilment may be available to human beings within it. In this regard it at least aligns human beings with the rest of the world, and with other life-forms, in an immensely long and fascinating story, the end of which is not yet in sight, and perhaps never will be. In this connection, Wilson as we saw, has emphasised the importance of a narrative in terms of which human beings can understand themselves as forming part of a larger whole. This is part and parcel of the search for explanation which has ultimately led to the Darwinian account itself.

Satisfying narratives, of course, have to take account of what are regarded as the important facts, as in the many Dreamtime stories of the Australian aboriginal peoples that explain the features of the Australian landscape of importance to these peoples, enable them to orient themselves with respect to them, and to transmit their self-understanding to their descendants for the latter's benefit. Science has found an unprecedentedly powerful way, not just of understanding the plainly observable facts, but also of unearthing hosts of previously undreamed-of new phenomena requiring explanation. Here is where the Darwinian narrative starts to become taxing, and probably faces serious difficulties as a substitute for the older tales. For it has stretched the account of life back further than human beings are probably evolved to take account of. We have great difficulty gaining any real appreciation of what a billion years is, whether in the past or future. It has unearthed facts and processes which are not easy to observe, and even harder to understand without a great deal of theoretical sophistication. In this regard it has precisely the same weaknesses as modern environmental concern. This, too, often involves the citing of threats to the environment that are hard to point out in a simple observational way. Also, understanding their nature requires a fairly extensive range of underlying facts and theories to be grasped before they become intelligible, let alone believable.

The consolations of this story, then, may take a lot of effort to access. Do such consolations exist? Does human life in general, and the life of individuals in particular, find a meaningful place within that story? The answer is, as ever, equivocal. On the one hand, the story often put forward at humanist funerals is still available within the Darwinian perspective, and perhaps gains added resonance from that perspective. This has it that individuals get their significance from the part they have played within the developing fabric of human life, however humble that part may be. They are a strand in its rich tapestry. The unspoken assumption is that their place in that fabric is guaranteed and cannot be undone, that their contribution will be always present even if the precise place of it is lost to view as time passes and memory fades.

The Darwinian perspective encourages us to think of ourselves as highly-interconnected with each other in this way, and indeed with all other life-forms. The tapestry extends to encompass the whole of life – as is so memorably delineated in Dawkins's account in *The Ancestor's Tale: A Pilgrimage to the Dawn of Life* (2004). But, as a story, it does have the features already alluded to. It has no overall point, no guiding intelligence and, being a scientific account it, avowedly, may be wrong. The temporal perspective involved is enormous, which may make it hard to grasp, though traditional religious ceremonies often refer to what has lasted and will last 'forever and ever'. But it is the indifference of the rest of the universe to human life that is likely to be the feature of the Darwinian account which most will find hard to accept as any form of consolatory story.

This, of course, is one of the key points of the parable with which this chapter began. It was also one of the reasons given in the last chapter for denying that Darwinism is, or could ever be, a new religion. It may, of course, be a 'post-religion' – a form of narrative which finally gives humanity a fully intelligible place, but no final point or purpose. It may be the main virtue of the Darwinian worldview, and of Edward O. Wilson's contribution to it, that it has allowed us to see that those two desiderata – intelligibility and purpose – are not at all the same, that only the former can be available to us and that perhaps we may learn satisfactorily to live in the light of that fact.

Appendix
The Basic Concepts of Neo-Darwinism and Sociobiology

Neo-Darwinism

(This section draws heavily on the masterly presentation of the current state of neo-Darwinism in Colin Patterson's *Evolution* (2nd edition 1999), especially Chapters 4, 5, 6 and 7.)

As every schoolchild knows, Charles Darwin's explanation of the origin of species claimed that the myriad species of flora and fauna that we see around us arose out of earlier forms by the process of natural selection. The key idea is that organisms are capable of producing many more offspring than in fact survive to reproduce themselves in turn. Not all offspring can survive, because the environment in which they live poses various threats, such as lack of food and water, predation, accident and so forth, which inevitably kill some before they can reproduce themselves. Those that do survive to reproduce must have coped with the environmental threats that destroyed their siblings.

There are always innate variations between members of a population that enable some to cope better than others with such threats. When those advantageous variations are passed on to offspring via the reproductive cells, then the favoured variants will spread through the population. Such favourable variations are called 'adaptations'. Given time, and the selective effects of differential environmental culling, new forms of organism will come into existence – new species will have been produced. As environments change, and/or as organisms move to meet new environments, this process of selection will continue, without limit, as will the emergence of new species and, through failures to adapt, the disappearance of existing ones.

The version of Darwin's ideas that has developed in the light of twentieth-century discoveries in genetics, which provided a clear and compelling picture of how organisms reproduce, is known as neo-Darwinism. Investigations into the microstructure of cells revealed the existence of the key elements involved in reproduction of organisms – the chromosomes – within the nucleus of each cell in an organism's body. Different species have different numbers of chromosomes, but they are always an even number, for in sexual reproduction half of them are passed on to offspring by each parent, via the egg or sperm, to make up a new full set of chromosomes. Eventually the chemical structure of these chromosomes, in the form of nucleic acids and proteins, was unravelled.

The proteins turned out to be the key components manipulated by the nucleic acids in cell metabolism. The nucleic acids, in the form of DNA and RNA, are the elements containing the genetic information – the genes – for the chemical processes of cell-building and metabolism, using the proteins in three ways – as catalysts (what are known as enzymes), structural materials and transporters. The famous Crick-Watson discovery of the double-helix structure of DNA finally revealed how this molecule could function in the storage and transmission of genetic information.

The molecule is like a twisted zip made out of a chain of alternating chemicals called 'bases' – adenine, guanine, cytosine and thymine. Each segment of the zip is made up of one of these bases 'locked on' to one of the other three as its partner – adenine always with thymine, and guanine always with cytosine. But the pairs of bases can occur in either order from left to right, and the sequence of the bases down the length of the zip is unrestricted.

The sequence of bases is thus a code for the building of various proteins that produce all the physical structure of the organism, including the complex brains of human beings and other mammals. The code is embodied in sequences of three bases – triplets. If the initial letter of the name of each base is used as a label (A,G,C,T), then these triplets can be expressed as sequences of three letters GAG, TCA, CGC and so forth. A code that is based on grouping the four bases into triplets gives 64 different possible combinations of letters, and thus 64 different possible 'statements' in the code. As it turns out, the code needs to specify only two things – amino acids, which are the building blocks of proteins; and a 'stop' sign, indicating where the message ends. There are only 21 possible amino acids, and so much of the code is not needed, with many of the triplets in fact having the same meaning and coding for the same amino acids.

Hence, organisms are built as the result of *triplets of DNA* coding for *amino acids* which form the *proteins* which act as structural components, transporters and enzymes within cells. Organisms are built out of these cells.

The information in the code is activated when the zip is undone. One side contains the coded message, which is read by another kind of nucleic acid, RNA. This is unlike DNA in being single-stranded and using one different base (uracil instead of thymine). The process of reading the code and converting it into proteins is complicated, and need not detain us. The other side of the zip is the template for the recreation of the code-bearing side of the zip, given the bonding rules already alluded to. Hence, both sides are needed – one to contain the code, the other to reproduce it.

This simple picture of the genes expressed in sequences of DNA has become very complicated by now. On the simple view, if a triplet of bases codes for one amino acid, and if we know how many amino acids are needed to make up a protein, then we can calculate the number of triplets to code for the whole protein. If we call the whole sequence of bases thus specified the 'gene' for that protein, then we appear to have specified precisely the molecular structure of that gene. But investigation revealed that actual genes for specific proteins contain many more stretches of bases than are needed for the coding. Such stretches of non-coding DNA are called 'introns'; the coding sequences are called 'exons'. The introns can be much longer than the exons, and thus genes seem to contain much redundant material that has to be excised precisely within the cell in the process of reading the code. In addition, another form of redundancy derives from the fact that identical coded messages may be repeated many times within the DNA molecule.

The activity of translating the code accurately into specific proteins is itself carried out by other specific proteins acting as enzymes. These too must be manufactured within the cell by the same process of decoding DNA sequences.

Since each cell of an organism's body contains the chromosomes, it follows that the code for creating the complete organism is contained within each cell. The term for the complete code is the 'genome'. But cells are differentiated into types, depending on which part of the organism's body they are going to construct. Hence, only part of the genome is activated within a particular type of cell, and the task of suppressing the unwanted genes within a given type of cell is again the work of specialist proteins, themselves produced by genes within the cells.

Representative human genomes have now been sequenced, containing some three billion base pairs. The number of genes within the human genome is very much less – about 30,000 – similar to the number in apparently simpler mammals, such as the mouse. But the number of genes is in itself not the key to understanding how genes produce organisms. What matters is how those genes are deployed by a master program within the genome. The same genes are deployed in different ways in different organisms, resulting in a dog rather than a rat, say. The genome is not a blueprint, where you can match an element in the blueprint with an element in the final construction. It is a recipe for producing various proteins in various sequences and assembling them in various ways.

The sequence of genes in the genome is essentially arbitrary, and so the construction of an organism based on the genetic code is not a matter of a chemical process reading the code in sequence from beginning to end along the genome and carrying out construction in that order. What is important is that the decoding and construction machinery of cells can find the appropriate genes when they are needed, wherever they are in the genome (Dawkins 2004: pp. 190–93).

It is important to guard against the idea, implicit in the erroneous 'blueprint' analogy, that genes are elements that control single features of the organism (known as the 'phenotype'). Rather, genes code for proteins, and these enter into complicated interactions which, taken together, may produce predictable effects on the final phenotype – its eye colour, say. But such regularities are always dependent upon the development's taking place within in certain typical environments. Change the latter, and the effects on the phenotype may be different, even though not random.

Sexual reproduction produces offspring that contain new versions of the genome. Each sexual cell, or gamete, contains only half of the parent's chromosomes, and when egg and sperm unite to begin producing a new phenotype there is a shuffling of the genes which produces variation. This variation, as we noted at the start, is crucial to the Darwinian account of the origin of species. This has suggested to many that we should conceive of separate species not simply in the traditional, biological, sense – as communities of organisms which can interbreed and produce fertile offspring – but as gene pools. A gene pool is 'the set of all genes in all the genomes of a sexually breeding population' (Dawkins 2004: p. 444). This implies that natural selection operates to enable some genes – those which produce traits which enhance the fitness of the phenotypes which possess them – to become more common in the gene pool, and others, which reduce fitness, to disappear from it. The gene pool thus changes over time, a difference marked by our referring to the evolution of new species and the disappearance of existing ones.

The genome is the source of the information needed to initiate cellular changes, by the mechanisms outlined above, but the received view is that there is no evidence that the process ever goes in the other direction – that information from the cell can bring about changes in the DNA. Hence, the only way that changes can be brought about in sequences of DNA by means of a regular mechanism is via the interaction of sexual reshuffling and natural selection operating on the resulting phenotypes. If DNA could be changed by cellular activity, then that would allow scope for the alternative hypothesis to account for evolution offered by Lamarck, who suggested characteristics acquired by an organism during the course of its life could be passed onto its offspring. The cell-to-DNA direction of change is not impossible, and recently some evidence has been found to suggest it may happen on at least some occasions with respect to the 'instruction manual' which directs which genes are switched on an off in various circumstances (see Vines 1998). But that in itself would not be sufficient to vindicate Lamarck against Darwin as a general theory.

The account just given of the way in which genes are reshuffled by sexual reproduction within a gene pool suggests that, identical siblings apart, each individual phenotype is unique, genetically-speaking. But we also noted above that the effect of a gene on the development of a phenotype is not usually a simple matter, and typically depends upon the interaction of the gene with the environment in which it finds itself – within the cell, within the body and within the ambient environment of the developing organism. In the case of human beings this will encompass the cultural as well as the physical environment – or, perhaps better, the physical environment as mediated and structured by cultural factors.

The perennially fraught question of the balance of influence of gene and environment on the development of the phenotype has long ceased to be a black and white issue amongst the proponents of neo-Darwinism. The consensus is that both are necessary, but to different degrees in different cases. Some phenotypic characters have an entirely genetic basis, such as the human ABO blood system.

Which blood group you are in – A, B, AB or O – depends entirely on which combination of genes from within a range of alternatives (called 'alleles') you have inherited from your parents. Other traits appear to vary much more widely, depending upon environmental factors (taking environment in the broadest sense already alluded to).

Human intelligence is the most disputed case, of course, given the crucial contribution of intelligence to the success and status of human beings amongst their own species. Twin studies, in which what are ex hypothesi genetically identical phenotypes that are raised from an early age in what are thought to be different environments are an obvious way forward. The twins are examined to detect significant variations between them with respect to various traits, including intelligence. If they end up with little variation in spite of very different environmental influences, then that appears to be strong evidence that variation in the traits in question is largely produced by genetic factors. But the results are rarely conclusive in settling the nature-nurture dispute, just because of the difficulty of specifying 'different environment' in such a way as to avoid all controversy. Perhaps it was the influence of the shared environment of the womb that was decisive, or a shared environment in the early post-natal phase of their lives; or perhaps there was sufficient similarity between their family milieux in key respects, in spite of differences in other respects, to account for the perceived (lack of) variation.

If variation between individuals is due largely to genetic factors then it is variation in the base sequences in the genome which is the cause. This, as we have seen, is standardly produced by the processes of sexual reproduction that shuffle the genes handed on by a set of parents to their offspring. However, another source of variation is more random, the result of an accident rather than the result of the normal workings of reproductive biology. Accidents in replication can result in changes to the sequence of DNA at some point – this is a called a point mutation. It can delete a letter, add one, or substitute a new one for an existing one. Not all point mutations produce an effect on the production of amino acids by DNA – it will be recollected that different triplets can code for the same amino acids and that some DNA sequences are introns, not involved in amino acid production anyway. But some point mutations will change the code in such a way that the letters are grouped into different triplets, coding for different amino acids and, thus, proteins – or sometimes resulting in failures to produce proteins at all.

The standard rate at which such mutations happen – as the result of the effects of natural radiation, say, or chemicals – has been calculated for a good range of organisms as being about one mutation per 100,000 genes per generation. Other forms of mutation also occur naturally, involving larger scale mutations at the level of chromosomes. These are less common than point mutations, however, which may have been the origin of the alleles already noticed – the variant genes which can be substituted at a given site upon the genome.

Many mutations have no detectable effect, others may be lethal and others may lead to variations in the phenotype which convey a reproductive advantage to their possessor, and thus may be the object of natural selection. However, what all this means is that, when neo-Darwinians develop their theory of natural selection in detail, they have to envisage it as acting upon the chromosome level and not just the single gene. They also need to note that genes, just because they are directly connected to adjacent genes in the chromosome, may become more common in a gene pool, not because they are directly selected, but because they are directly connected to another gene that is.

The question of whether or not a mutation that does have an effect on its bearer – the phenotype – is harmful or beneficial is always a relative matter. In the theory of natural selection, the key idea is relative reproductive success. A mutation is beneficial if it increases the chance that an organism will leave fertile offspring, as compared to its rivals. This does not mean, of course, that it is a 'good thing' in any moral, or other humanly important, sense, that improved reproductive success of some genotypes as compared with that of others takes place, or that evolution produces better or worse states of affairs. Benefits and harms are here given a technical sense relating simply to the effects of certain mutations upon the content of the gene pool.

From the 'gene's eye view', of course, what is a 'benefit' or 'harm' to it is whether or not it is spreading or diminishing in the gene pool. But genes are just stretches of DNA. They do not literally have a point of view or a sense of self, and so are only metaphorically called 'selfish'. They do not, because they cannot, actively and knowingly seek such expansion in the gene pool. To spread in the gene pool, they have to produce effects, probably in conjunction with other genes, and via interaction with the various aspects of their environment, upon the ability of their carrier to engage in successful reproductive behaviour. That way they will be duplicated (the individual stretch of DNA does not survive, rather copies of itself are made down the generations) and their future copies will have the same effect on their bearers in turn.

There is, nevertheless, a genuinely disquieting idea operating here, which is that there is now no need, for explanatory purposes, to suppose that any organisms, including human beings, have come into existence as the result of some mind's desire that there should be such things. They are not the realisation of some conscious purpose. It is just that organisms, in specific environments, with those features which have been causally produced by the actions of genes, have reproduced those genes successfully, more successfully than their competitors and thus have continued to exist, and multiply, with those features. But there is nothing in this explanation which grants any such set of organisms a privileged status, or guarantees that a gene which has successfully expanded in its gene pool in the past will continue to do so.

A 'successful' gene, in this sense, resulting from a mutation is necessarily a lucky gene. Most mutations that have an effect upon their phenotypes, bestowing upon them a variant trait that affects their reproductive success, will almost inevitably harm them. That is because, given the very workings of natural selection, most of the time organisms will be well adapted to their environments. However, a changing environment, or the encounter with a new one, can rapidly change the picture.

As noted earlier, it is no longer plausible to think of genes as coding directly for specific features of organisms. In most cases a single gene is only a necessary condition for the development of that feature in the phenotype, and as many theorists have been arguing, it may well not be the only necessary condition, or the only hereditary condition needed for the feature to develop. This seems to make the neo-Darwinian picture we have been considering a rather complex one. But it is also clear that it is implausible to suppose that every feature of an organism is present because it improves the organism's reproductive fitness. Even if we suppose that some of these features are side-effects (exaptations) of genes that are selected for their contributions to fitness, and so can be brought into the theory of natural selection at one remove, it has become clear that there are many variant features of organisms – at the chemical level of the cell, for example – that are impossible to think of as adaptations in the Darwinian sense. Variations in the DNA sequences in introns, for example, can have no obvious value for reproductive success. These are dead or psuedo-genes. Hence, we have the recent development of the alternative theory that at the level of DNA organisms are subject to a great deal of random genetic drift – changes in DNA which are the result of chance.

However, this idea does not in itself challenge the idea of natural selection. It rather suggests that the scope of the latter is less than first appeared, not that it is any the less important where it does apply. Contrariwise, neo-Darwinism has also become relaxed about the idea that features of organisms can exist and persist for reasons that have nothing to do with natural selection. But natural selection remains the only viable scientific theory to explain the origin of species.

Sociobiology

(This section draws heavily on Chapters 2 and 5 of Edward O.Wilson's *Sociobiology: The Abridged Edition* (1980).)

This form of enquiry takes the neo-Darwinian account just given for granted, and uses its guiding concepts to investigate the biological basis of the social behaviour of those organisms that can correctly be characterised as inhabiting societies. A society for these purposes may be defined as 'a group of individuals belonging to the same species and organised in a cooperative manner' (Wilson 1980: p. 7).

Societies of this kind can be found across the living world, from invertebrates, such as ants, to vertebrates, such as primates and, of course, to human beings. Social organisation is understood as an emergent property which depends on the interaction of individual organisms within groups, but which cannot be reduced to an analysis of individual behaviour. Societies, that is, are wholes possessing distinctive properties that have to be conceptualised separately from those which apply to their individual members. Sociobiology seeks to develop a set of concepts and theories that will allow the identification of the common properties of societies across all such instances and to use the neo-Darwinian theory of natural selection to cast light upon their evolution. That is, societies, just as much as organisms, are taken to be entities that can properly be understood to have an evolutionary history and be analysed with the aid of the concept of natural selection.

The aim of sociobiology is, then, to be able to predict the forms of social organisation for a given species once one knows key facts about it. Such facts encompass, for example, how the genetic make-up of that species constrains the behaviour of its members and the demographics of the society, which reveal facts about its population growth and age structure.

The essential phenomenon which makes individual members of a species into members of the same society is that they engage in 'reciprocal communication of a cooperative nature, transcending mere sexual activity' (Wilson 1980: p. 7). This enables us to delimit a society in terms of the existence of boundaries across which such communication is much less frequent. Communication has to be understood sufficiently broadly to allow use of the same concept for organisms which communicate by means of a language in the human sense and those which use other means. Hence, it is defined in terms of an 'action on the part of one organism (or cell) that alters the probability pattern of behaviour in another organism (or cell) in an adaptive fashion' (Wilson 1980: p. 9).

The concept of a society differs from the key concept of evolutionary biology, namely that of a 'population' of organisms. A population is a geographically delimited set of organisms of the same species, and is thus characterised by the capacity of its members to interbreed freely. Its boundaries thus mark out a line across which genes, rather than communication, flows less freely. Often society and population coincide, as when social relationships inhibit the exchange of genes across the social boundaries, while permitting gene flow across subgroups of the society (Wilson 1980: pp. 8–9). But the two may not be identical, as when communication flows across boundaries that are delimited in terms of gene flow.

Although sociobiology seeks a genetic basis for the understanding of societies, it recognises that the level of social organisation is the furthest removed from the genetic level, with intermediate processes operating so as to convert genetic changes into, sometimes dramatic, social transformations. This is the 'multiplier effect' whereby small changes in societies, or small differences between societies, at the level of individual behaviour explain large-scale changes/differences at the social level. Socialisation practices are an example of these kinds of influences (Wilson 1980: p. 9).

The idea that a society can possess an evolutionary history, and be subject to natural selection, implies that we can analyse the components of social structure that are adaptations and use them to assess the society's overall degree of fitness.

A crucial factor of this kind will often be the society's demography. Societies contain differentiated individuals, such as male and female, young and old, parent and child and perhaps caste differences – such as' worker' and 'queen' in the social insects. A society which is well-adjusted to its environment will contain some appropriate relative proportions of each of these categories, and its overall behaviour as a society will be determined by them. Clearly, a society containing only elderly males faces early extinction, and thus societies which are going to avoid this fate need the appropriate, and usually stable, age distribution in order to produce new members, raise them to maturity, and encourage/permit them to reproduce.

However, while some societies are so structured that their demographic profile is vital to their fitness, and thus can properly be regarded as an adaptation, this cannot be assumed always to be the case. Sometimes what is selected is certain kinds of individual behaviour, with the demographic profile produced by this a mere side effect, with no influence on the fitness of the society as a whole. This may apply, for example, to opportunistic species that achieve fitness by adopting the strategy of producing large numbers of offspring as quickly as possible when circumstances permit. This tends to produce a flattened age pyramid, resulting from lots of offspring but a short life span. Such a distribution does not itself contribute anything to the fitness of the population (Wilson 1980: p. 11).

Other factors besides demographic structure constitute key elements in the analysis of sociality. *Group size* (where a group is a set of conspecifics that remain in close contact with each other and interact more frequently than with other conspecifics) is one such factor: the sizes of different kinds of groups within various societies exhibit certain regularities. *Cohesiveness* – the degree of physical closeness of group members to each other – is another key variable. Then, it is important whether the *interconnection* between members of groups is patterned or unpatterned. That is, acts of communication may be preferentially directed by individuals to only certain other categories of individual (as with hierarchies) or be directed indiscriminately to whoever happens to be around. Societies may differ in terms of their degree of *permeability*, that is the extent to which they permit immigrants or communicate with other societies of the same species within their vicinity. Different societies show different degrees of *compartmentalisation*, which concerns how easily they form, or separate into, separate sub-units – such as family groups. Degree of role *differentiation* is another crucial variable, for fitness can be improved by the specialisation of individuals in certain roles, making possible the division of labour. But then it becomes important how much *integration* of specialised behaviour characterises a society with differentiation of roles. Such integration in turn will require *information flow*, specified in terms of number of signals, information content per signal, and rate of information flow. Finally, sociality can be gauged by the *amount of time devoted to social activity*, which again shows marked variation across social species (Wilson 1980: pp. 12–14).

It is crucial when sociobiologists attempt to characterise the social characteristics of some species by using these concepts that they take account of 'behavioural scaling'. This is the idea that social behaviour in a species may alter, sometimes drastically, in a predictable way depending upon the state of certain key variables which are impinging upon members of specific populations – such as population density. Thus, aggressive tendencies may be non-operative at low densities, but increasingly manifest themselves as population increases. Thus, social behaviour in a species may be scalar in form – different modes of behaviour are elicited to different degrees in response to changes in the degree of some stimulus. What should be regarded as a possible adaptation, therefore, is behaviour across the scale, which may be regarded as fitting the organism for the whole variety of circumstances within which it may find itself. Description of the social characteristics of members of the species will thus be rendered complex by this factor (Wilson 1980: pp. 14–15).

There is, however, one key feature of sociality that has for long seemed to pose a fundamental challenge to any Darwinian account of societies, and that is the existence of altruism. Wilson, indeed, suggests that the 'central theoretical problem of sociobiology' is the question of how altruism, which apparently reduces the fitness of an organism by inducing it to put the interests of another, or others, ahead of its own can evolve by natural selection (Wilson 1980: p. 3). The answer, first hinted at by Darwin himself, was developed by William Hamilton in the 1960s and 1970s (Hamilton 1964). The fitness of individuals is measured by the number of surviving offspring. But if we adopt the gene's eye view, then offspring are simply bearers of genes. In producing offspring individuals are passing on their genes to a new generation. But, crucially, a proportion of an individual's genes may be present in conspecifics other than offspring, such as siblings, cousins and so forth.

If individuals behave altruistically, putting the interests of other individuals who share their genes ahead of their own interests, they may nevertheless be helping to pass on their genes to the next generation, thereby preserving, or extending, the representation of those genes in the gene pool. If the tendency to altruism is itself genetically based, then that very gene is likely to be reproduced also, furthering the spread of the tendency to altruism throughout the gene pool.

To express this thought Hamilton coined the idea of 'inclusive fitness', which covers the fitness of an individual (genes passed on directly by that individual to offspring) plus the parts of that fitness (fractions of the individual's genes passed on) to be found in all related individuals. Hence, genetically based altruism will evolve if it increases the inclusive fitness of individuals within groups which display it, as compared with groups which do not.

Another key element of sociality, especially amongst human beings, is reciprocity – doing favours to others in the expectation that they will return the favour in future. The key Darwinian theorist of this kind of behaviour was Robert Trivers (1971), who argued that individual fitness, defined as above, is enhanced in any society in which 'reciprocal altruism' of this kind becomes prevalent. The explanation of why those who benefit from favours do not cheat when their time comes to reciprocate is that amongst species, such as human beings, in which individuals are highly personalised and thus are identified and remembered as individuals, cheats tend to be punished. They thereby tend to lose any advantage in personal fitness with which their cheating initially provided them. Trivers has argued that many of the human tendencies to moral attitudes, such as moral outrage, gratitude, sincerity and guilt have evolved as a means of maintaining the institutions of reciprocal altruism, so central to human life, against the depradations of cheats.

These two explanations of the evolution of key elements of sociality, above all of human sociality, may or may not be successful. However, they are crucial for the sociobiological explanation of morality and thus, given the intermingling of morality and religion, play a large part in the sociobiological understanding of the role of religion in human life.

Bibliography

Agar, N. (1996), 'Teleology and Genes', *Biology and Philosophy*, 11: 289–300.

Axelrod, R. and Hamilton, W. (1981), 'The Evolution of Cooperation', *Science*, 211: 390–96.

Barkow, J. (1992), 'Beneath New Culture is Old Psychology: Gossip and Social Stratification', in J. Barkow, L. Cosmides and J. Tooby (eds), *The Adapted Mind: Evolutionary Psychology and the Generation of Culture* (New York: Oxford University Press).

Baxter, B. (1999), *Ecologism: An Introduction* (Edinburgh: Edinburgh University Press).

Baxter, B. (2004), *A Theory of Ecological Justice* (London: Routledge).

Benton, T. (1991), 'Biology and Social Science: Why the Return of the Repressed Should be Given a (Cautious) Welcome', *Sociology*, 25: 1–29.

Bowlby, J. (1969), *Attachment and Loss. Vol 1: Attachment* (London: Hogarth Press).

Boyd, R. and Richerson, P.J. (1985), *Culture and the Evolutionary Process* (Chicago: University of Chicago Press).

Brown, D. (1991), *Human Universals* (New York: McGraw-Hill).

Bryson, B. (2003), *A Short History of Nearly Everything* (London: Doubleday).

Buss, D.M. (1999), *Evolutionary Psychology: The New Science of the Mind* (London: Allyn and Bacon).

Callicott, J.B. (1989), *In Defense of the Land Ethic: Essays in Environmental Philosophy* (Albany, NY: SUNY Press).

Cavalli-Sforza, L. and Feldman, M. (1981), *Cultural Transmission and Evolution: A Quantitative Approach* (Princeton, NJ: Princeton University Press).

Christus Rex Information Service (2006), online: http://www.christusrex.org/www1/pope/ (accessed11 May 2006).

Connelly, J. and Smith, G. (1999), *Politics of the Environment: From Theory to Practice* (London: Routledge).

Count, E.W. (1958), 'The Biological Basis of Human Sociality', *American Anthropologist*, 60 (6): 1049–85.

Cosmides, L. and Tooby, J. (1992), 'Cognitive Adaptations for Social Exchange', in J. Barkow, L. Cosmides and J. Tooby (eds), *The Adapted Mind: Evolutionary Psychology and the Generation of Culture* (New York: Oxford University Press).

Darwin, C. (1901), *The Descent of Man and Selection in Relation to Sex* (2nd edn), (London: John Murray).

Dawkins, R. (1976), *The Selfish Gene* (Oxford: Oxford University Press).

Dawkins, R. (2004), *The Ancestor's Tale: A Pilgrimage to the Dawn of Life* (London: Weidenfeld and Nicholson).

de Waal, F. and Lanting, F. (1997), *Bonobo: The Forgotten Ape* (Berkeley, CA: University of California Press).

Dennett, D. (1995), *Darwin's Dangerous Idea: Evolution and the Meanings of Life* (London: Allen Lane/The Penguin Press).

Dobzhansky, T. (1963), 'Anthropology and the Natural Sciences – the Problem of Human Evolution', *Current Anthropology*, 4 (138): 146–8.

Dupré, J. (2001), *Human Nature and the Limits of Science* (Oxford: Oxford University Press).

Durkheim, E. (1982), *The Rules of the Sociological Method*, trans. W.D. Halls, ed. S. Lukes (Basingstoke: Macmillan).

Endler, J. (1986), *Natural Selection in the Wild* (Princeton, NJ: Princeton University Press).

Feyerabend, P. (1975), *Against Method* (London: Verso).

Gare, A. (1995), *Postmodernism and the Environmental Crisis* (London and New York: Routledge).

Goodpaster, K. (1978), 'On Being Morally Considerable', *Journal of Philosophy*, 75: 308–25.

Gould, S. (1989), *Wonderful Life: The Burgess Shale and the Nature of History* (New York: Norton).

Gould, S. and Lewontin, R. (1979), 'The Spandrels of San Marco and the Panglossian Paradigm: A Critique of the Adaptationist Programme', *Proceedings of the Royal Society*, B205: 581–98.

Hacker, P. (1972), *Insight and Illusion: Wittgenstein on Philosophy and the Metaphysics of Experience* (Oxford: Oxford University Press).

Hamilton, W. (1964), 'The Evolution of Social Behaviour', *Journal of Theoretical Biology*, 7: 1–32.

Harvey, P. and Pagel, M. (1991), *The Comparative Method in Evolutionary Biology* (Oxford: Oxford University Press).

Hargrove, E. (1989), *Foundations of Environmental Ethics* (Denton, TX: Environmental Ethics Books).

Hayek, F. (1949), *Individualism and Economic Order* (London: Routledge).

Hayek, F. (1960), *The Constitution of Liberty* (London: Routledge and Kegan Paul.

Heywood, A. (2003), *Political Ideologies: An Introduction* (3rd edn), (London: Palgrave).

Hinchman, L. (2004), 'Is Environmentalism a Humanism?', *Environmental Values*, 13: 3–29.

Kant, I. (1958), *Groundwork of the Metaphysic of Morals*, trans. H. Paton (New York: HarperTorch).

Kant, I. (1993), *Critique of Practical Reason*, trans. Lewis Beck (New York: Macmillan/Library of Liberal Arts).

Krebs, J. and Davies, N. (1987), *An Introduction to Behavioural Ecology* (2nd edn), (Boston, MA: Blackwell Scientific Publications).

Laland, K. and Brown, G. (2002), *Sense and Nonsense: Evolutionary Perspectives on Human Behaviour* (Oxford: Oxford University Press).

Lumsden, C. and Wilson, E.O. (1981), *Genes, Minds and culture: The Co-evolutionary Process* (Cambridge, MA: Harvard University Press).

Lynch, T. and Wells, D. (1998), 'Non-anthropocentrism? A Killing Objection', *Environmental Values*, 7 (2): 151–63.

MacArthur, R. and Wilson, E.O. (1967), *The Theory of Island Biogeography* (Princeton, NJ: Princeton University Press).

Maynard Smith, J. and Price, G. (1973), 'The Logic of Animal Conflict', *Nature*, 246: 15–18.

Marsh, D. and Stoker, G. (eds) (2002), *Theory and Methods in Political Science* (2nd edn) (Houndmills and New York: Palgrave Macmillan).

Maslow, A. (1972), *The Farther Reaches of Human Nature* (New York: Viking Press).

McEwan, I. (2006), 'A Parallel Tradition', *The Guardian*, *Review* section, pp. 4–6, 1 April.

Midgley, M. (1979), *Beast and Man: The Roots of Human Nature* (Brighton: Harvester Press).

Midgley, M. (1994), *The Ethical Primate: Humans, Freedom and Morality* (London and New York: Routledge).

Midgley, M. (2002), *Evolution as a Religion* (2nd edn) (London and New York: Routledge).

Miller, D. (1989), *Market, State and Community: Theoretical Foundations of Market Socialism* (Oxford: Clarendon Press).

Norton, B. (1991), *Toward Unity among Environmentalists* (Oxford and New York: Oxford University Press).

Orzack, S. and Sober, E. (2001), 'Adaptation, Phylogenetic Inertia and the Method of Controlled Comparisons', in S. Orzack and E. Sober (eds), *Adaptationism and Optimality* (Cambridge: Cambridge University Press).

Oyama, S. (1985), *The Ontogeny of Information* (Cambridge: Cambridge University Press).

Partridge, E. (1984), 'Nature as Moral Resource', *Environmental Ethics*, 6: 101–30.

Patterson, C. (1999), *Evolution* (2nd edn) (London: Natural History Museum).

Pinker, S. (1997), *How the Mind Works* (London: Penguin Books).

Plumwood, V. (1993), *Feminism and the Mastery of Nature* (London: Routledge).

Rachels, J. (1990), *Created from Animals: The Moral Implications of Darwinism* (Oxford: Oxford University Press).

Richards, J. Radcliffe (2000), *Human Nature after Darwin: A Philosophical Introduction* (London: Routledge).

Richerson, P. and Boyd, R. (1998), 'The Evolution of Human Ultra-sociality', in I. Eibl-Eibesfeldt and F. Salter (eds), *Indoctrinability, Warfare and Ideology: Evolutionary Perspectives* (Oxford: Berghahn Books).

Ridley, M. (1993), *The Red Queen: Sex and the Evolution of Human Nature* (New York: Macmillan).

Ridley, M. (1996), *The Origins of Virtue* (London: Viking).

Ridley, M. (2003), *Nature via Nurture: Genes, Experience and What Makes Us Human* (London: HarperCollins (Harper Perennial)).

Ruse, M. (1986), *Taking Darwin Seriously: A Naturalistic Approach to Philosophy* (Oxford: Blackwell).

Schelling, T. (1978), *Micromotives and Macrobehaviour* (New York: W.W. Norton.

Sideris, L. (2003), *Environmental Ethics, Ecological Theology and Natural Selection* (New York and Chichester: Columbia University Press).

Sinervo, B. and Basolo, A. (1996), 'Testing Adaptation using Phenotypic Manipulations', in M. Rose and G. Lauder (eds), *Adaptation* (San Diego, CA: Academic Press).

Smith, M. (2001), *An Ethics of Place* (Albany, NY: SUNY Press).

Sterelny, K. and Griffiths, P. (1999), *Sex and Death: An Introduction to the Philosophy of Biology* (Chicago and London: University of Chicago Press).

Sterelny, K., Smith, K. and Dickison, M. (1996), 'The Extended Replicator', *Biology and Philosophy*, 11: 377–403.

Strawson, P. (1959), *Individuals* (London: Methuen).

Symons, D. (1992), 'On the Use and Misuse of Darwinism in the Study of Human Behaviour', in J. Barkow, L. Cosmides and J. Tooby (eds), *The Adapted Mind: Evolutionary Psychology and the Generation of Culture* (New York: Oxford University Press).

Taylor, C. (1991), *The Ethics of Authenticity* (Cambridge: Cambridge University Press).

Tiger, L. and Fox, R. (1971), *The Imperial Animal* (New York: Holt, Rinehart and Wilson.

Tooby J. and Cosmides, L. (1990), 'The Past Explains the Present: Emotional Adaptations and the Structure of Ancestral Environments', *Ethology and Sociobiology*, 11: 375–424.

Tooby, J. and Cosmides, L. (1992), 'The Psychological Foundations of Culture', in J. Barkow, L. Cosmides and J. Tooby (eds), *The Adapted Mind: Evolutionary Psychology and the Generation of Culture* (New York: Oxford University Press).

Trivers, R. (1971), 'The Evolution of Reciprocal Altruism', *Quarterly Review of Biology*, 46: 35–57.

Trivers, R. (1972), 'Parental Investment and Sexual Selection', in B.Campbell (ed.), *Sexual Selection and the Descent of Man, 1871–1971* (Chicago: Aldine).

Trivers, R. (1974), 'Parent–Offspring Conflict', *American Zoologist*, 14: 249–64.

Vines, G. (1998), 'Hidden Inheritance', *New Scientist*, 2162: 27–30.

Wagner, G. (ed.) (2001), *The Character Concept in Evolutionary Biology* (San Diego, CA: Academic Press).

Walzer, M. (1983), *Spheres of Justice: a Defence of Pluralism and Equality* (Oxford: Robertson).

Weiner, A. (1978), 'Epistemology and Ethnographic Reality: A Trobriand Island Case Study', *American Anthropologist*, 80: 752–57.

Wilson, E.O. (1975), *Sociobiology: The New Synthesis* (Cambridge, MA and London: Belknap Press).

Wilson, E.O. (1978), *On Human Nature* (Cambridge, MA : Harvard University Press).

Wilson, E.O. (1980), *Sociobiology: The Abridged Edition* (Cambridge, MA and London: Belknap Press).

Wilson, E.O. (1984), *Biophilia* (Cambridge, MA: Harvard University Press).

Wilson, E.O. (1992), *The Diversity of Life* (Cambridge, MA: The Belknap Press).

Wilson, E.O. (1996), *Naturalist* (London: Penguin Books).

Wilson, E.O. (1998), *Consilience: The Unity of Knowledge* (London: Little, Brown and Co. (UK) (Abacus)).

Wilson, E.O. (2002), *The Future of Life* (London: Little, Brown and Co.).

Winch, P. (1958), *The Idea of a Social Science* (London: Routledge and Kegan Paul).

Wittgenstein, L. (1953), *Philosophical Investigations* (Oxford: Blackwell).

Wolpert, L. (2006), *Six Impossible Things before Breakfast* (London: Faber and Faber).

Wright, R. (1995), *The Moral Animal: Evolutionary Psychology and Everyday Life* (London: Little, Brown and Co.).

Index